TRIBOLOGY IN ENGINEERING

Edited by **Haşim Pihtili**

INTECHOPEN.COM

Tribology in Engineering
http://dx.doi.org/10.5772/3305
Edited by Haşim Pihtili

Contributors

S.O. Yılmaz, M. Aksoy, C. Ozel, H. Pıhtılı, M. Gür, Erkan Bahçe, Cihan Ozel, Ibrahim Sevim, Andrzej Kulczycki, Czesław Kajdas, Juan R. Laguna-Camacho, M. Vite-Torres, E.A. Gallardo-Hernández, E.E. Vera-Cárdenas, G. Belforte, F. Colombo, T. Raparelli, A. Trivella, V. Viktorov, Qinling Bi, Shengyu Zhu, Weimin Liu, M. Shahabuddin, H.H. Masjuki, M.A. Kalam, H. Allmaier, C. Priestner, D.E. Sander, F.M. Reich, Irfan Çelal Engin, Czesław Kajdas

Published by InTech

Janeza Trdine 9, 51000 Rijeka, Croatia

Edition 2014

Publishing Process Manager Marina Jozipovic
Technical Editor InTech DTP team
Cover InTech Design team

Additional hard copies can be obtained from orders@intechopen.com

Tribology in Engineering, Edited by Haşim Pihtili
p. cm.
ISBN 978-953-51-1126-9

INTECH

open science | open minds

Contents

Preface

Although the study of friction and wear caught the attention of many eminent scientists during the course of the past few centuries, consistent and sustained scientific investigation into friction and wear is a relatively recent phenomenon. Tribology is therefore a comparatively young science where rigorous analytical concepts have not yet been established to provide a clear guide to the complex characteristics of wear and friction. Much of the tribological research is applied or commercially orientated and already a wide range of wear-resistant or friction-reducing materials have been developed. The basic concept of tribology is that friction and wear are best controlled with a thin layer or intervening film of material separating sliding, rolling, and impacting bodies. There is almost no restriction on the type of material that can form such a film and some solids, liquids and gases are equally effective. If no film material is supplied then the process of wear itself may generate a substitute film. The aim of tribology is either to find the optimum film material for a given application or to predict the sequence of events when a sliding, rolling, impacting contact is left to generate its own intervening film.

Chapters in this book describe the fundamental wear mechanisms operating in metallic and non- metallic materials together with some prognoses concerning the future developments of these materials. Two classes of materials with entirely different characteristics polymers and ceramics are discussed. Polymers can provide low friction and low wear coefficients but their use is limited to lower temperatures and consequently low speeds and loads. Ceramics are resistant to high temperatures and often have a good wear resistance but their applications are limited by poor friction coefficients, especially in unlubricated applications. Ceramics and polymers are surprisingly vulnerable to accelerated wear in the presence of corrosive reagents and care should be taken in the selection of materials that are appropriate for particular operating conditions. None of these materials meet current or future needs, and much effort is being expended to develop new materials and improve the properties of existing materials for new and future applications. The development of polymer and ceramic matrix composites reinforced by fibres, platelets, and particulates serves as an example of these efforts.

Very few tribology chapters exist and these are dated. They have focused on conventional tribology, traditional materials and matured industrial applications. The purpose of this book is to present the principles of tribology, mechanics and applications, to various applications. The appeal of the subject book is expected to be broad. The chapters in the book have been written by internationally recognized experts in the field, from academia, national research labs and industry, and from all over the world. The book integrates the knowledge of the field from mechanics and materials science points of view. It covers various measurement techniques and their applications, and theoretical modelling of interfaces. Organization of the book is straightforward. The parts of the book cover fundamental experimental and theoretical studies on the tribological and friction applications caused by many operating machine parts. The book is intended for three types of readers: graduate students of tribology, mechanics, nanotechnology, research workers who are active or intend to become active in this field, and practicing engineers who have encountered a tribology and mechanics problem and hope to solve it as expeditiously as possible.

Asst. Prof. Dr. Haşim Pihtili
Mechanical Engineering Department
Fırat University,
Elazig,
Turkey

Section 1

Wear

Chapter 1

Effect of FeCr Intermetallic on Wear Resistance of Fe-Based Composites

S.O. Yılmaz, M. Aksoy, C. Ozel, H. Pıhtılı and M. Gür

Additional information is available at the end of the chapter

http://dx.doi.org/10.5772/55869

1. Introduction

Metal-matrix composites (MMCs) have higher stiffness and mechanical strength than alloys, however they have lower ductility and fracture toughness [1]. In microstructure of MMCs if a bond between particulate reinforcement and matrix has been constituted, then the composite exhibits an ability to withstand high tensile and compressive stresses. Continuous fibers, short or chopped fibers, whiskers and particulates have been used as reinforcement materials in MMCs. Discontinuous reinforcement phase composites are common due to availability, low cost, independence of mechanical properties from particulate orientation [2] and production via a vide range of manufacturing routes [3-6].

Wear is described as the removal of material from a surface in relative motion by mechanical or chemical processes [7]. The wear of the materials can be formed due to adhesion, abrasion, surface fatigue or tribochemical reaction [8,9]. The removal of material from the surface by hard particles (three-body abrasion) or by a rough counter face (two-body abrasion) is generally termed as abrasive wear. The wear resistance of a material is related to its microstructure, and the changes in microstructure may take place during the wear process [10,11] Developments of lightweight and energy-saving materials have become more numerous in the past few years in many different fields [12-16]. Recent studies [17-25] indicated that a significant improvement in the tribological properties of Fe alloys can be attained by the addition of hard carbides. Metallurgical processing, such as casting and powder metallurgy (P/M) techniques, has been successfully employed to produce anti-abrasion Fe-based composites consisting of hard carbide particles [17-25]. The strength of the as-cast composites is usually less than that of the P/M composites, and it is also possible that some large casting defects exist in the cast. These problems can largely be overcome in the P/M route. Additional advantages of the P/M process are that a high dislocation density can be introduced into the matrix, recrystalization can be prevented by carbide reinforcements, and in their structure their subgrain size is small.

Most studies [17-21] indicated that the wear resistance of MMCSs manufactured by the P/M and/or casting techniques increased with increasing volume fraction of reinforcement particulates. The wear resistance of the composite decreased with increasing reinforcement above a certain level. Jha *et al.* [26] indicated that the wear rates increased with increasing reinforcement volume fraction in the P/M sintered soft matrix alloys.

In this study, the Fe base P/M composites are reinforced with FeCr carbide complexes, soft graphite, and Cu particles to improve wear resistance benefiting from the advantage of both the energy absorption properties of the soft matrix phases and the wear resistance of the hard carbide phases. With this aim, we investigated the microstructures, wear properties and some mechanical properties (surface hardness, tensile strength and toughness of Fe base MMCSs) by using scanning electron microscopy (SEM), surface hardness (HB), tensile testing, Charpy V-notch impact and abrasive wear tests.

2. Experimental procedure

The chemical compositions of the FeCr particulates are given in Table 1. Composites containing 1 to 15wt% Fe/Cr particulates with an average particle size of about 20 μm were prepared by a conventional P/M process, which involved the steps of mixing, cold isostatic pressing (1000 MPa), degassing and sintering according to the schedules in Table 2.

The hardness of the samples were measured in the range of ±3 error band with HB hardness scale under 612.5 N load. In addition, the toughness of the samples was evaluated in the range of ±0.6 error band using Charpy V-notch specimens. The tensile strength test samples were prepared upon ASTM E8-78 L_0= 4d standard [27] and the tests were performed under Hounsfield type machine at room temperature, and by using a crosshead speed of 50 mm/min.

A pin-on disk apparatus was used for evaluating the abrasive wear resistance. For the abrasive wear tests, cylindrical billets of 12.5 mm diameter and 10 mm height were machined. Before the wear tests, each specimen was ground up to grade 1200 abrasive paper, making sure that the wear surface completely contacted the surface of the abrasive paper. Abrasive wear tests were carried out under dry sliding conditions by sliding the sample under an applied load of 10, 20, 30 and 40 N respectively over a grade 80 abrasive paper stuck to the grinding disk, which rotted at 320 rev min^{-1}. A fixed track diameter of 160 mm was used in all tests, and the duration of abrading was 60 s. Each test was conducted using a fresh abrasive paper. For each test condition, at least three runs were performed. Wear rates were obtained by determining the weight loss of the samples before and after wear tests.

Samples for microscopic examination were prepared by standard metaleographic procedures; they were then etched with %1 nital reagent and examined by optical and scanning electron microscopy (SEM). For determination of the wear mechanism of the Fe alloy and its composites, the worn surfaces and debris were examined by scanning electron microscopy, where the samples were gold coated prior to examinations.

	Element amount (wt %)					
	Cr	Fe	Si	C	P	S
Fe/Cr	64	26,30	1,80	6,84	0,02	0,038

Table 1. The chemical compositions of the Fe/Cr particulates

Treatment	Temperature (^{0}C)	Time (h)	Atmosphere	Remarks
Sintering	900	1	Ar	Degassing
Sintering	1200	2	Ar	Liquid phase sintering
Solutionizing	1100	2	Ar	As solutionized

Table 2. Processing schedules of the powder metallurgy Fe composites.

3. Results and discussions

3.1. Microsturucture

The microstructures of the composites with FeCr reinforcement were investigated and optical micrograph of the sample S_1 is given in Figure 1. It was seen that, the microstructure of the FeCr reinforced MMCSs consist of ferrite matrix with dispersed FeCr particulates. The addition of graphite to the composite with FeCr particulates formed different phases (Figure

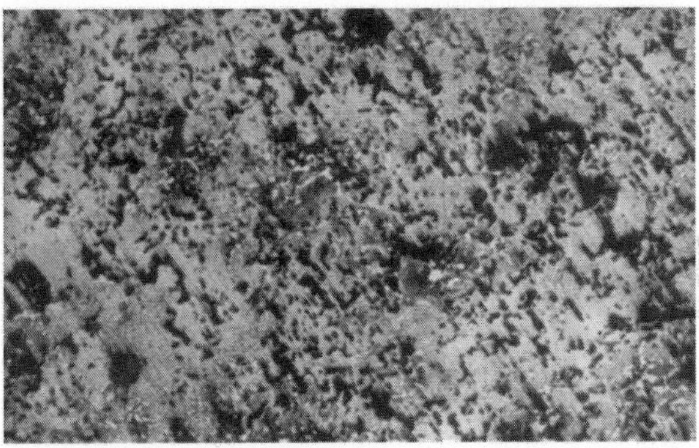

Figure 1. Optical micrographs of the sample S_1

2-Table 3). Depending on graphite amount, pearlite phase started to form around graphite particles, and increasing the amount of graphite increased the ratio of pearlite phase and M_3C carbides. Graphite grains were formed in the pearlite structure in samples with 1wt% graphite supplement. On the other hand, by increasing graphite content to 2wt%, ledeburitic structure has been formed in grain boundaries besides formation of M_3C carbides at grain boundaries and toward center of grains (Figure 3). The microstructures of the samples having soft copper supplement in the range 0.5-2 wt% with graphite (0.5wt%) and FeCr (5wt%) particulates were found near to each other, but their microstructures was different than the samples having a structure without copper supplement (Figure 4-5 and Table 4.).

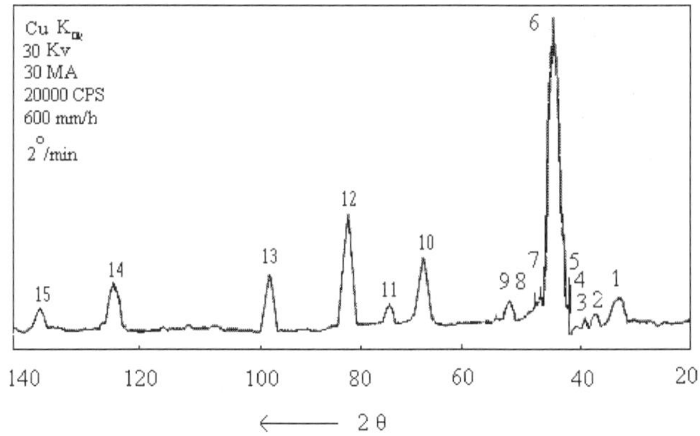

Figure 2. X-ray difractom of the sample S_{10}

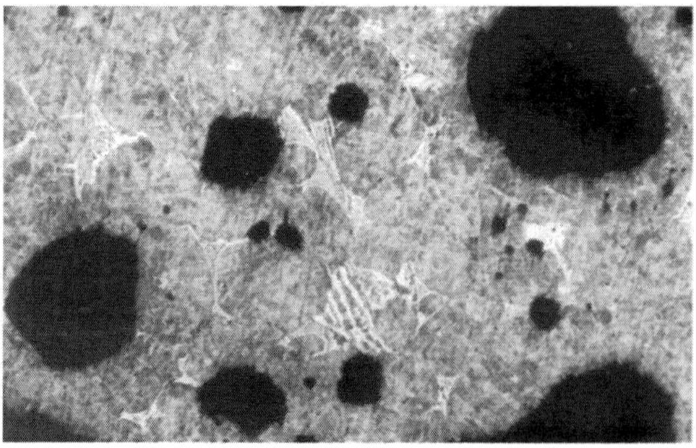

Figure 3. Optical micrographs of the sample S_{10}.

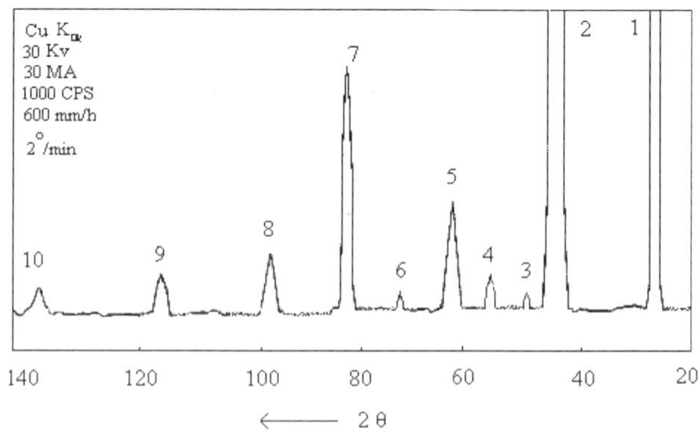

Figure 4. X-ray difractom of the samples S_{13}

Phases		Pikes														
		1	2	3	4	5	6	7	8	9	10	11	12	13	14	15
M_3C	S8		200				201									
	S9	101	200	110			201									
	S10	101	200		002	210	201		111	110		210				
M_7C_3	S8		211	413	101											
	S9	222	211		413	101										
	S10		222				211				413		101			
αFe	S8				110	200	211	220	310					222		
	S9		110		200	211	220	310	222							
	S10						110				200		211	220	310	222
C	S8			002		110										
	S9		002	100		110										
	S10				002		100					110				
$M_{23}C_6$	S8		111	200			220									222
	S9	111	200			222										
	S10			111			200						220		311	222

Table 3. X ray pikes of sample S_{10}

Phases	Pikes									
	1	2	3	4	5	6	7	8	9	10
M_7C_3			100			102	110			
MC_3		020		201	011	110				
αFe	111	110		200	211	220	210	222		
$M_{23}C_6$		111		200		220	211		310	222
C	002				110			300		400

Table 4. X ray pikes of S_{13}

3.2. Mechanical testing

The results of surface hardness, toughness, tensile strength and wear resistance tests are summarized in Table 5. The hardness of the FeCr reinforced samples (S_1-S_6) increased by FeCr reinforcement (Figure 6.a). The reason for this increase in hardness can be attributed to the increase in wt% of FeCr reinforcement and the diffusion of dissolved Cr atoms into the matrix. Moreover, the addition of graphite with FeCr particulates increased hardness significantly (Figure 6.b). The reason is thought to be due to formation of pearlite phase in matrix, and as well due to the ledeburite, M_3C_2, and σ phases (Figure 4-Table 3). In addition to the X-ray diffraction, EDS analysis form these phases are taken and it was detected as; the amount of Cr is 6wt.% and 8-12 wt.% in M_3C and ledeburitic structure, respectively. Furthermore, the hardness data of the samples S_7-S_{10} also showed well agreement with the presence of the phases obtained (Table 3).

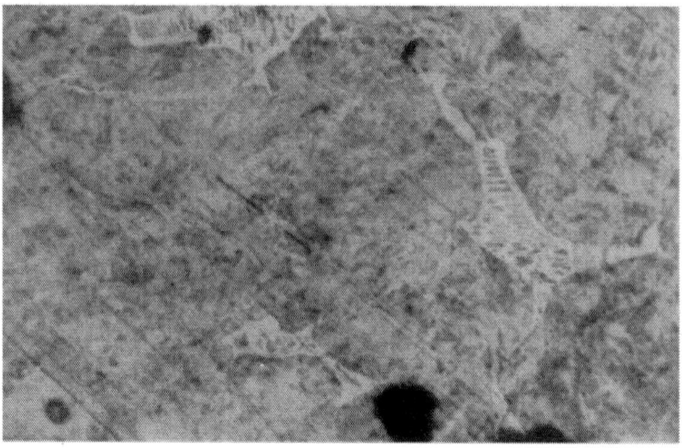

Figure 5. Optical micrographs of the sample S13

	Composition	Hardness HB	Toughness J Cm²	T Strength MPa	Weight Loss (mg) 10 N	20 N	30 N
S1	Fe	55,75	486	27	0,57	0,87	1,12
S2	Fe+1wt%FeCr	64,4	336	24	0,55	0,83	1,09
S3	Fe+3wt%FeCr	72	384	32	0,6	1,93	1,235
S4	Fe+5wt%feCr	77,66	207,5	41,5	0,7	1,023	1,254
S5	Fe+7wt%FeCr	94	48	48	0,95	1,1	1,51
S6	Fe+10wt%FeCr	116	40,95	58,5	1,2	1,4	1,859
S7	Fe+5wt%FeCr+ 0,25wt% Graphite	79	492,64	61,58	0,78	0,87	1,12
S8	Fe+5wt%FeCr+ 0,5wt% Graphite	95	504	72	0,61	0,75	0,98
S9	Fe+5wt%FeCr+ 1wt% Graphite	121,75	420	84	0,57	0,7	0,94
S10	Fe+5wt%FeCr+ 2wt% Graphite	281	311,5	89	0,18	0,27	0,31
S11	Fe+5wt%FeCr+0,5wt%Graphite+%0,5wt%Cu	214	429	78	0,31	0,45	0,5
S12	Fe+5wt%FeCr+0,5wt%Graphite+1wt%Cu	220	346,4	86,6	0,29	0,4	0,45
S13	Fe+5wt%FeCr+0,5wt%Graphite+2wt%Cu	224	241,975	96,79	0,21	0,34	0,4

Table 5. The mean values and error bands of results of surface hardness, toughness and wear resistance tests

The effect of copper supplement with graphite and FeCr particulates on the hardness of the samples was detected for the samples namely S_{11}, S_{12} and S_{13}. The relationship between hardness and wear resistance was given in Figure 6.c as function of Cu concentration. As seen from the figure, Cu supplement increased hardness considerably. The increase in hardness is probably due to the microstructural change, and the reason for formation of martensitic and bainitic structures are due to presence of alloying elements. C, Cr and Cu elements in the structure decrease formation temperature of martensite (Ms) and bainite (Bs) and they provoke formation of these phases [28,29]. It was detected that, copper is present as 0.05-0.25, 1.43-2.87, 1.98-4.94 wt% in martensitic+bainitic zone, respectively, and Cr is detected in the range of 1.98-4.94 wt% Cr in the same samples by EDS analysis. In other words, the amount of Cr in martensitic and bainitic structures increased with increasing the amount of copper supplement.

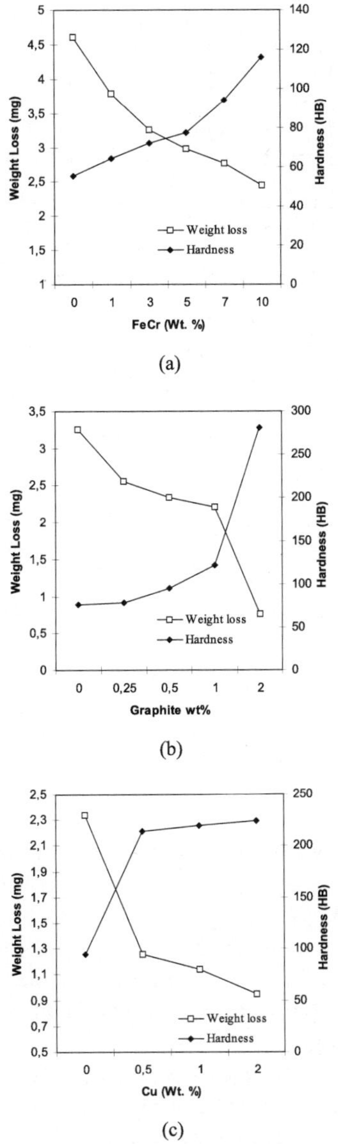

Figure 6. The relationship between hardness and weight loss as function of (a) FeCr (b) graphite (c) copper concentration

Reinforcement of the FeCr carbides increased toughness of the samples (S_1-S_6) significantly (Figure 7.a). However, the toughness of the samples having graphite was changed, and an optimum point for the amount of graphite was found. At the beginning, graphite increased toughness, however after 0.5 wt.% graphite additions, the amount of graphite decreased toughness (Figure 7.b). This decrease is attributed to the presence of graphite particles, hard intermetallic phases and the diffusion of Cr into the matrix producing a brittle structure.

The weight loss of the samples S_7, S_8, S_9 and S_{10} decreased with the increase in the amount of graphite. Investigations on microstructure of the samples having graphite supplement in the range 0.25 to 2 wt% (with 5wt% FeCr particulates) have showed that additional phases were formed. Moreover, the increase in the amount of graphite also decreased the size of FeCr particulates. It was conjectured that the decrease of particulate size improved wear resistance. Because, the good bonding between the composite constituents avoids third body abrasion and allow to FeCr particulates to act as load-bearing elements of the composite. Investigations on the toughness of the samples having copper show that copper addition decreased toughness (Figure 7.c), because copper increased the amount of Cr in all of the phases, and provided formation of martensite and bainite phases. Furthermore it increased the diffusion rate and decreased the size of carbides

The change in weight loss vs. load of the samples S_1-S_6 is given in Figure 8.a. For all loads, the highest weight loss obtained for the sample S_6 and the sample S_1 gave the lowest weight loss. Investigations on the microstructure of the sample S_6 show that, the matrix of S_6 was constituted from ductile ferrite phase. Hence, it is conjectured that FeCr particulates were easily pulled out during wear. Also, within craters (*i.e.*, where flakes of material cracked and wore away) FeCr particulates protruded from the surface. Particularly the sample S_{10} has shown the lowest weight loss and a different wear behavior. On the other hand for all loads the wear rate of the samples S_6-S_9 are near to each other (Figure 8.b).

Copper was added to the matrix of the samples S_{11}-S_{13} having FeCr particulates (5wt.%) and graphite (0.5wt.%) together to decrease amount of porosity and friction coefficient. It was observed that copper increased hardness, tensile strength, but decreased toughness and weight loss. The change of weight loss with load has given in Figure 8.c. From the figure it is seen that the effect of load decreased with copper supplement. The relationship between hardness and wear resistance of the samples are given in Figure 9 for 30 N load. The wear tests show that there isn't correlations between wear resistance and hardness for the samples (S_1-S_6) having FeCr reinforcements under abrasive tests over 80 grade abrasives. As the hardness increased, the weight loss increased. On the other hand, addition of graphite to the matrix of the samples (S_7-S_{10}) increased hardness and decreased weight loss, but toughness of the samples didn't changed parallel to the weight loss of the samples. Because, the microstructure of the matrix has been changed by diffusion of carbon atoms into the matrix, and besides shrinkage of carbide particulates have been seen. Furthermore, the Cu supplements have shown that the wear rate of samples S_{11}, S_{12}, S_{13} changed proportionally with surface hardness.

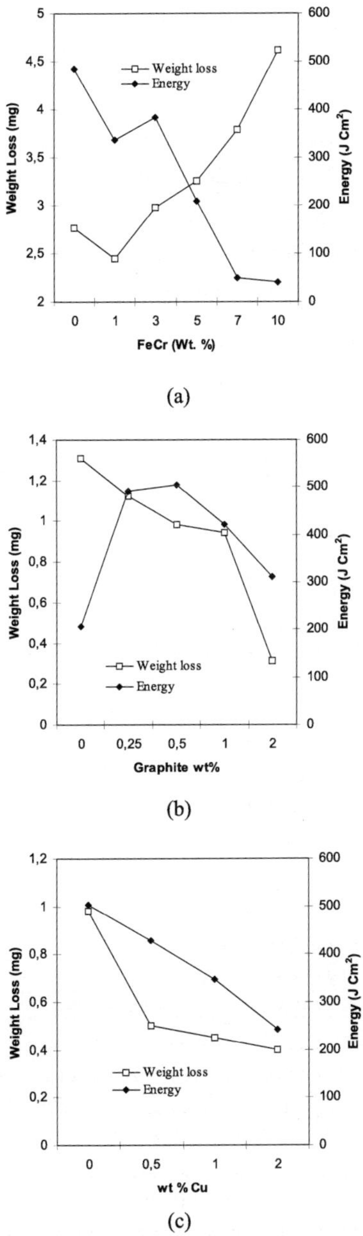

Figure 7. The relationship between toughness and weight loss as function of (a) FeCr (b) graphite (c) copper concentration

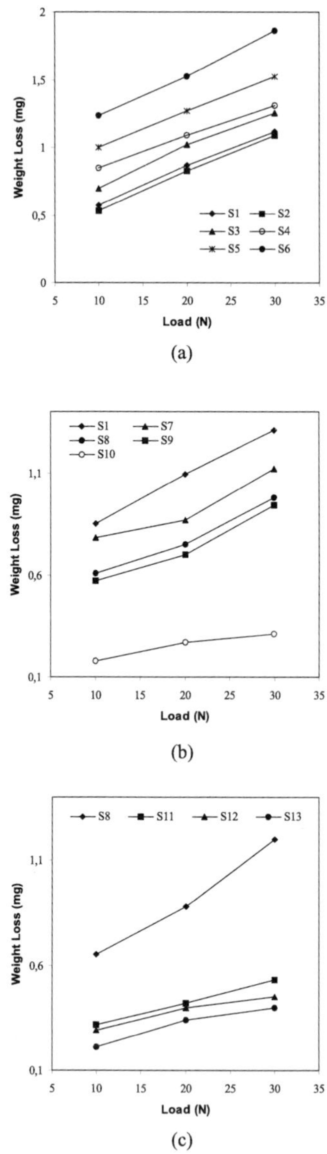

Figure 8. Wear rate vs. load for the samples (a) S_1-S_6 (b) S_7-S_{10} (c) S_{11}-S_{13}

A qualitative difference was found between tracks of Fe/FeCr/graphite-Fe/FeCr/Cu and Fe/FeCr specimens, suggesting less sensitivity to the load of the former. This delay of the load effect can be attributed to the presence of FeCr particulates, size of particulates and hardness of the matrix, and particulates indicated that they act as load-bearing elements more efficiently than ceramics. Furthermore hard abrasive ceramic reinforcements, such as SiC, have the deleterious effect of wearing the counterface more than the unreinforced material does [30,31]. In addition, as cracks might propagate through the matrix/ceramic reinforcement interface [32,30] pulled out ceramic particles may act as third-body abrasion elements [32] of both specimens and counterfaces, worsening wear behavior of composite/counterface system.

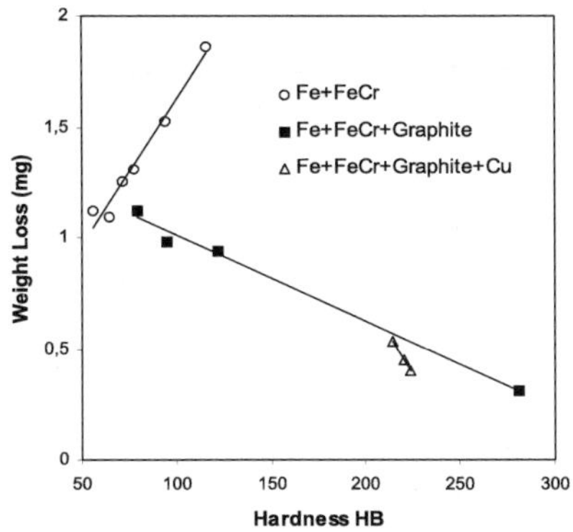

Figure 9. The relationship between hardness and weight loss (30 N load and 80 grade abrasive).

4. Conclusion

The wear tests applied over 80 grade abrasive papers have shown that the weight loss of the MMCs having only FeCr particulates increased with increase of FeCr ratio. However, increase in wt.% of FeCr particulates increased hardness, toughness and yield strength. Graphite supplement with FeCr particulates have formed additional phases, decreased size of FeCr particulates and increased matrix hardness. Hence, weight loss decreased, and increasing graphite amount increased hardness linearly. On the other hand, toughness of the samples having graphite additives decreased after 0.5wt.% graphite. Nevertheless, the wear rate of the samples were changed accordingly to load, but the wear rate of S_{10} didn't changed in a considerable amount with load, and its rate was far low than other samples. Samples with copper additive have shown an increase in hardness, tensile strength with Cu amount. However, toughness was decreased with weight loss. In addition the dependence of the weight loss of the samples with copper to the load decreased with copper addition.

Author details

S.O. Yılmaz* and M. Aksoy
Department of Material and Metallurgical Engineering Fırat University Elaziğ, Turkey

* Corresponding Author

C. Ozel, H. Pıhtılı and M. Gür
Department of Mecahnaical Engineering Fırat University Elaziğ, Turkey

5. References

[1] Metal powder Industries Federation: "*Powder metallurgy Equipment Manuel*" New York, 1968.

[2] D.L. McDanels, *Metall. Trans.*16A (1985) 1105

[3] C. Millliere and M. Suery, *J. Mater.Sci.* 34-36 (1988) 41.

[4] C.G.Levi, G.J. Abbascian and R. Mehrabian, *Metall, Trans.*, 9A (1978) 697.

[5] A. Mortensen and I. Jin, *Int. Mater. Rev.*, 37 (3) (1992) 101.

[6] F.M. Hosking, F.F. Portillo, R. Wunderlin and R. Mehrabin, *J. Mater.Sci.*, 17 (1982) 477.

[7] Ovingsbo, Proc. Conf. Wear of Materials, ASME, New York, 1979, p.620.

[8] K.H. Zum Gahr, Microstructure and Wear of Materials, Elsevier, Amsterdam, 1987, pp.132-495.

[9] D.Godfrey, Diagnosis of Wear Mechanism, in: M.B. Peterson, W.O. Winer (Eds.), Wear Control Handbook, ASME, New York, 1980, pp. 283-312.

[10] Y. Wang, T.C. Lei and C.Q. Gao, *Tribol. Int.*, 23(1) (1980) 47-53.

[11] D.A. Rigney and W.A. Glaeser, Wear of Materials, ASME, New York. 1977, pp. 41-46.

[12] T.L. Ho, M.B. Peterson. F.F. Ling, *Wear* 30 (1974) 73-91.

[13] T.L. Ho, , M.B. Peterson.. *Wear* 43 (1977) 199-210.

[14] T.L. Ho, in: W.A. Glasser (E.), Wear of Materials, American Society of Mechanical engineers, New york, 1977, pp.70-76.

[15] R. Munro, International Congress and Exposition, Detroit, MI, 28 February-4 march 1983, SAE Tech Paper 830067.

[16] J. Dinwoodie, E. Moore, C.A.J. Langman, W.R. Symes Jr, in: W.C. Harrigan. J. Strife, J. Metallurgical Society of AIME. Metalurgiacal Society of AIME, Warrendale, P.A. 1985, pp. 671-685.

[17] H.L. Lee, W.H. Lu and S.L.I. Chan, Chin., *J. Mater.Sci.*, 24 (1) (1992) 40-52

[18] D.L. Erich, Int. *J. Powder Metal.*, 23 (1) (1987) 45-54

[19] S.V. Prasad and P.K. Rohatgi, *J. Met. Sci.*, 39 (11) (1987) 22-26

[20] K.J. Bhansali and R. mehrabian, *J. Met. Sci.*, 34 (9) (1982) 30-34

[21] F.M. Hosking, F. F. Portillo, R. Wunderlin and R. mehrabian, *J. Mater. Sci.*, 17 (1982) 477-498.

[22] M.K. Surappa, S. V. Prasad and P.K. Rohatgi, *Wear,* 77 (1982) 295-302.

[23] P.R. Gibson, A.J. Cleg and A. A. Das, *Wear,* 95 (1984) 193-198.

[24] S.J. Lin and K.S. Liu, *Wear,* 121 (1988) 1-14.

[25] Y.M. Pan, M.E. Fine and H.S. Cheng, *Scr. Metal.*, 24 (1990) 1341-1345.

[26] A.K. Jha, S.V.Pprasad and G.S. Upadhyaya, *Wear*, 133 (1989) 163-172.

[27] E.A Brandes, Metals Reference Book. 6 th edition. Butterworths, London. (1983).

[28] A. Wang and H.J. Rack. *Mater. Sci. Eng.*, (1991), vol. A147, pp.211-24

[29] I.M. Hutchings: *Mater. Sci. Eng.*, (1994), A184, pp. 185-95.

[30] A.T. Alpas and J. Zhang: *Metall. Mater. Trans.* A, (1994), vol. 184, pp. 187-92.

[31] W.Ames and A.T. Alpas: *Metall. Mater. Trans.* A, (1995), vol. 26A, pp. 85-98

[32] P.L. Ratnaparkhi and H.J. Rack: *Mater. Sci. Eng.*, (1990), vol. A129, pp. 11-19.

Chapter 2

Experimental Investigation of the Effect of Machining Parameters on the Surface Roughness and the Formation of Built Up Edge (BUE) in the Drilling of Al 5005

Erkan Bahçe and Cihan Ozel

Additional information is available at the end of the chapter

http://dx.doi.org/10.5772/56027

1. Introduction

Aluminum is one of the metals, besides iron and steel, which are widely used in many industrial fields such as aviation, navigation and automotive. The most considerable reasons why aluminum is so widely used are: i) it is light, ii) its alloys have higher strength than the construction steel, iii) it has high heat and electrical conductivity. Because of these excellent characteristics, the usage of aluminum as engineering material has an ever-increasing importance in several technological fields [1,2]. Although aluminum is widely used, there are many problems such as tool abrasion, burr formation and poor hole surface quality in drilling process of aluminum and its alloys [2,3]

The surface quality, which is one of these problems, is quite important for the efficient working of machine parts. The structure of a machined surface is one of the most important criteria in terms of quality, and tribological properties of the machined surface are considerably affected from the surface tissue. Generally the surface quality is characterized with surface roughness. Surface roughness is an important factor which must be considered not only in the conventional subjects of tribology such as abrasion, friction and lubrication but also in different fields such as sealing, hydrodynamics, electrical and heat conductivity. Surface roughness is mainly affected during the machining process by cutting parameters such as cutting speed, feed rate and depth of cut [4,5,6]. If these parameters are not chosen convenient, the surface roughness increases. This situation creates a notch effect and results in crack initiation, decrease in fatigue strength and corrosion resistance. So, the characterization and measurement of surface roughness has a great important in the sense of the optimization of machining process [7,8].

In former studies on hole surface quality, Nouari and his colleagues subjected Al 2024-T3 to dry drilling process, and did some optimizations and analysis experimentally for both the dimensional accuracy of the machined surface and longevity of cutting tools [9]. In their study, they used sintered tungsten carbide (STC) cutting tools and high speed stell (HSS) cutting tools, and set the feed rate to 0,04 mm/rev and cutting speed to 25, 65, 165 m/min. They concluded from the experiments that STC cutting tools are more convenient in comparison with HSS cutting tools from the points of tool life, deviation in hole diameter and surface roughness. Lin investigated tool life, surface roughness, tool abrasion and burr formation for the process of the high speed machining of stainless steel material with TiN coated carbide tool [10]. As a result of his researches, he determined that the abrasions in shear edge result from the high feed rate in low cutting speed, and optimum cutting speed for desired burr height and surface roughness was 75 m/min. In addition, he determined that in high speed machining of stainless steels the tool life increased considerably in case of adjusting the feed rate to the values lower than 0,05 mm/rev. Lin and Syhu, studied on the treatment of the tool life and burr formation in the drilling of stainless steel with the drill bits coated by different materials [11]. Kurt et al., investigated the effect of cutting parameters on the drilling temperature, cutting force and surface roughness in the drilling of Al 2024 alloy with DLC coated drill. In their study, they determined that the most effective factors influencing the hole surface quality are feed rate and drill diameter [12]. They observed that the change in feed rate and diameter at high cutting speeds affects the average surface roughness considerably. Dudzinski et al., determined that the tool life was very short in the drilling of Inconel 718; therefore the surface quality gets worse [13]. They determined that the main wear mechanism seen in the cutting tools used was abrasion. In addition, they observed that the chips resulted in the formation of built-up-edge (BUE) by adhering on the cutting tool, and the removal of BUE from the cutting tool repeatedly caused notches. Kılıçkap investigated the roughness of hole surface and the height of the burrs formed at the hole exit in the drilling of Al 7075 material [14]. Also in another research, Kılıçkap, experimentally studied on the effects of cutting speed, feed rate and different cooling techniques on the temperature and the roughness of hole surface in the drilling of Al 7075 [15]. In their study, they observed that the most appropriate cooling technique was oil cooling from the point of good surface roughness. Also, they determined that the roughness increased with the increase of the feed rate, while it decreased with the increase of rotation speed. Hanyu et al. investigated the effects of finely crystallized diamond coating method, which was developed by themselves, on the surface roughness in the dry and semi-dry drilling of Al 7075 alloy [16]. They demonstrated experimentally that finely crystallized diamond coating method yields four times better results in comparison with the conventional diamond coating method. Konig and Grass investigated the effects of cutting parameters on the roughness of hole surface and surface tissue in the drilling of fiber reinforced thermosets [17]. They denoted that the surface roughness increases with increase of the feed rate. In his study, Tosun, optimized the drilling parameters affecting the burr height and surface roughness of DIN 42CrMo4 steel material by considering different drill materials, cutting speeds, drill point angles and feed rates with the help of Grey Relational Analysis (GRA) [18]. Sur et al. studied on the effects of Ti alloy on the surface roughness in the turning of Al 6063 alloy [19]. They observed that the increase of 35 percent in the hardness of the material resulting from the doping of Ti to the material had relatively an inconsiderable effect on the surface roughness of the material in comparison with effects of cutting speed and feed rate. Also, they determined that the increase in the feed rate affected the surface roughness negatively, while the increase in the cutting speed contributed to the treatment of surface roughness positively; however the feed rate had a more dominant effect on the surface roughness in comparison with the cutting speed. Darwish, et al., investigated the effects of cutting speed, feed rate and drill diameter on the hole surface quality, dimensional accuracy and geometric tolerance in soft steel materials [20]. In their study, they observed that cutting speed and feed rate had a great effect on surface quality, and the higher dimensional accuracy was obtained at low cutting speeds and feed rates.

In the studies mentioned above, generally the effects of cutting parameters on the roughness of the hole surface were investigated in the machining process of stainless steel and 2000, 6000 and 7000 series aluminum alloys. However, it has drawn attention that the studies on 5000 series aluminum alloys, which are widely used in many industrial fields such as aviation, navigation and automotive, are not sufficient. In this study, Al 5005 material was drilled on CNC milling machine under dry drilling conditions by considering different machining parameters such as various rotation speeds, feed rates, drill diameters and point angles, and the roughness of hole surface and the formation of BUE on cutting edges were investigated.

2. Experimental method

In this study, Al 5005 was drilled by considering various drilling parameters such as diameter, point angle, feed rate and rotation speed. CNC milling machine (Taksan, TMC 700V) with vertical machining centre was used in the experiments. The spindle power of the machine, rotation speed and feed rate values were taken as 5.5 kW, 50-8000 rev/min and maximum 0.6 mm/rev, respectively. Maximum feed rate values of the work table on X, Y and Z axes were 500, 600 and 450 mm, respectively. Factorial design, in which the effects of mostly different and unrelated factors on a definite characteristic are investigated, was taken into consideration in design process of the experiment. In factorial design, the experimental design is established by processing the variable parameters (or their levels) crossingly [21]. In this study, the experiments were conducted in accordance with 72 different combinations ($2^1.3^2.4^1$) by using 2 levels for the drill diameter, 4 levels for the point angle, 3 levels for the rotation speed and the feed rate. The values of variable parameters in conducted experiment were selected in compliance with the similar studies as shown in Table 1 [9,10,14,18].

In this study, the cutting fluid was not used in order to observe the effect of drill parameters on the roughness of the hole surface [22]. Al 5005 material used in the experiments was in the dimension of 10mmx70mmx400mm, and its chemical properties were given in Table 2. In the drilling process, the space between the axes of each hole on the sample was adjusted to be 20 mm (Figure 1).

Feed Rate (mm/rev)	Rotation Speed (rev/min)	Point Angle (degree)	Drill material and diameter (mm)
0.1, 0.2, 0.3	400, 800, 1200	90, 118, 130,140	HSS, Ø5, Ø10

Table 1. Experimental parameters

Al 5005	Mg	Si	Fe	Cu	Mn	Cr	Zn	Other elements	Al
%	0.5-1.10	0.3	0.3	0.2	0.20	0.10	0.25	0.15	remainder

Table 2. The chemical structure of Al 5005

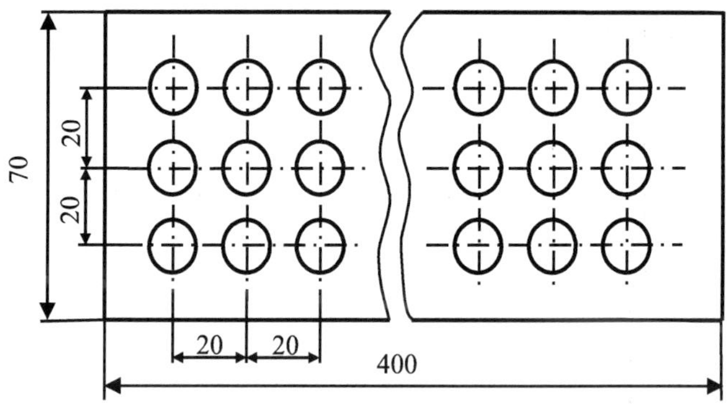

Figure 1. The space between the drilling axes (Thickness 10 mm)

N type double-end DIN 338/RN HSS drill bits with 30º helix angle were used in drilling process. Hardness value of these cutting tools was 65 HRc. Each cutting tool was used once in the experiments, and each experiment was repeated three times in accordance with the similar studies [9,11-14,21,22]. The images of BUEs formed on the cutting tool as a result of the drilling processes were taken by means of Leo Evo 40 model Scanning Electron Microscope (SEM).

After combinational drilling processes, the samples were cut with a cutting disc in the middle in parallel with the hole axis in order to measure the roughness of the hole surface. Then, the surface roughnesses were measured with Mitutoyo SJ-201 surface roughness measurement device. In the measurement of the roughness, sampling length and sampling number were chosen by considering the former studies [7,8,16,19,21] as 0.8 mm and 5 (0.8x5), respectively. The other sampling length values of this device were 0.25 mm and 2.5 mm. Generally, the roughness was measured at three different points in parallel with the hole axis in accordance with the studies in literature [7,12,16,18]. But in this study, in order to evaluate the measurements accurately, the measurements were taken from 5 different points, and then Ra values were determined by considering the average of these values.

3. Results and discussion

The graphics in Figure 2 were illustrated to enable one a comprehensive assessment of the effects of drilling parameters on the surface roughness in the drilling of Al 5005 without using cooling fluid.

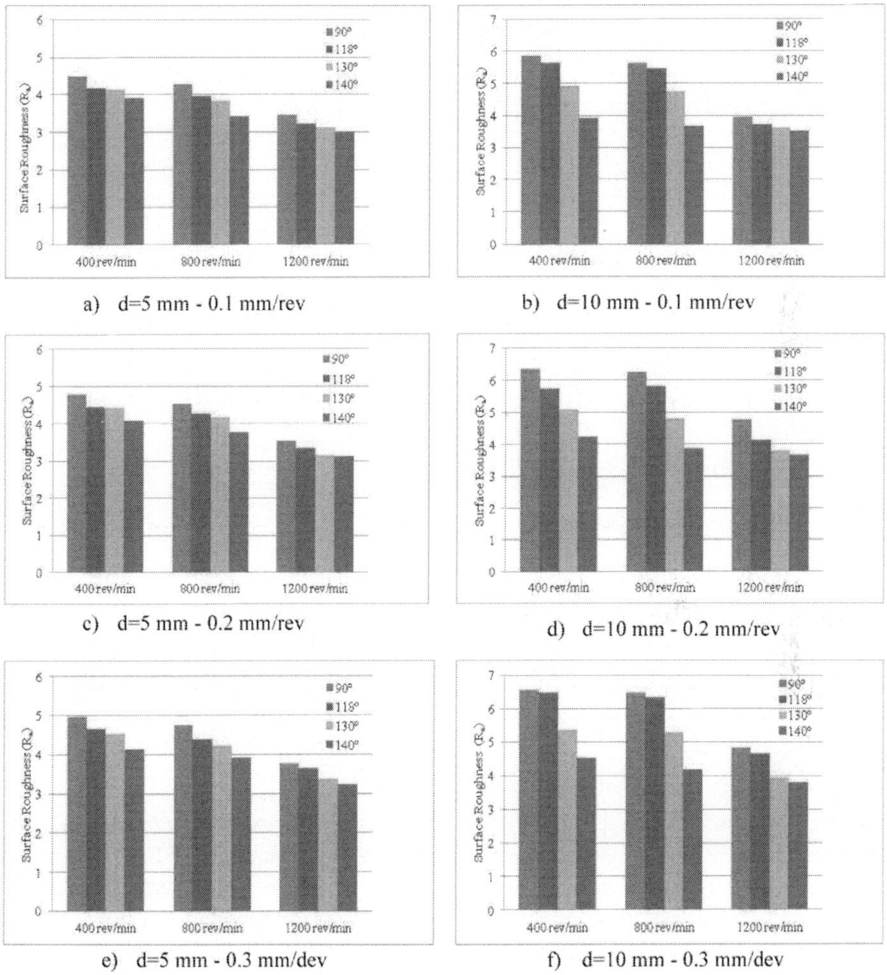

Figure 2. The change of the surface roughness with the drilling parameters

As seen from the graphics, the surface roughness decreases with the increase of rotation speed. In former studies [18,23], this case was attributed to the decrease in the cutting and feed force. The most considerable reasons of the decrease in these forces are the decrease in the contact region between the tool and work piece and the decrease in the shear strength in the cutting region due to the increase in the heat of tool-work piece depending on the cutting speed [18,24].

Also, it was supposed that the influence of BUE on the tool and material increased negatively since the amount of BUE resulting from the adhesive wear increased due to the increase in the feed and cutting forces. As mentioned in the literature, in the machining processes of many alloys including more than one phase in own structures BUE formed due to the adhesion of the chips on the tool surface and cutting edge because of the work hardening [23]. It can be said that BUE especially forming at low cutting speeds affects the surface roughness negatively. since Aluminum and its alloys includes more than one phase. In order to explain this case clearly, BUE formations occurring on the cutting tool edges were also investigated in this study (Figures 3 and 4 a-c). All the images corresponding to the whole cutting parameters were not presented since there were a lot of parameters in the experiment and they were investigated in other sections separately, but the SEM images expressing the mentioned case clearly were presented.

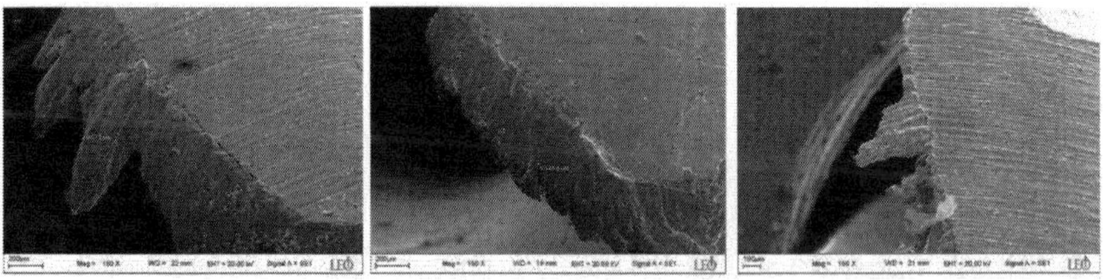

a) 400 rev/min b) 800 rev/min c) 1200 rev/min

Figure 3. SEM images of BUE formation on the cutting edges. (0.2 mm/dev-118°- Ø5 mm)

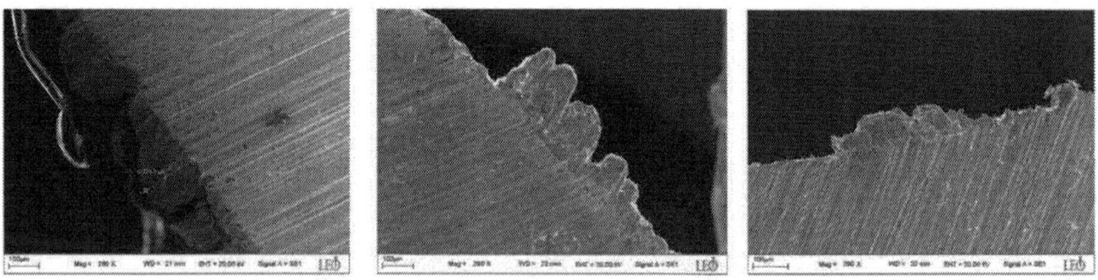

a) 400 rev/min b) 800 rev/min c) 1200 rev/min

Figure 4. SEM images of BUE formation on the cutting edges. (0.1 mm/rev-118°-Ø10 mm)

As seen from the SEM images in Figures 3 and 4 a-c, BUE formation decreased and had a minor effect on the surface roughness because of the increase in the rotation speed, so surface roughness decreased (Figures 4a-f). Since BUE formed on the cutting tool edge during the drilling had an unstable structure, the surface roughness increased. Thus, because of big and unstable BUE due to low cutting speeds (Figures 3, 4 a and b), the surface roughness increased further and a bad surface was formed (Figures 2 a-f). The decrease in BUE due to the increase in cutting speed can be ascribed to the increase in temperature [23,25]. Since high cutting speeds resulted in much more increase in the temperature, BUE on the cutting edge lost its hardness and strength, and in the continuing cutting process it couldn't resist the tensions on itself and it was removed from the cutting edge (Figure 3 and 4 c). Hence, high cutting speeds reduced the tendency to the formation of BUE, and resulted in the decrease in the surface roughness values of the work piece (Figures 2 a-f). Since BUE formed on the cutting edges also spoiled the geometric structure of the cutting tool, the stable and ideal process of the cutting operation was damaged, so the roughness increased (Figure 2 a-f). Also, BUE formed on the cutting edges caused fracturing and abrasion on the cutting edges while it was separating from the cutting edges by the effect of thermal tensions [24]. This case increased the roughness of the hole surface depending on the size of BUE (Figures 4 a-c). The abrasion and fracturing formed on the cutting edge according to the different machining parameters were presented in Figures 5 and 6, respectively.

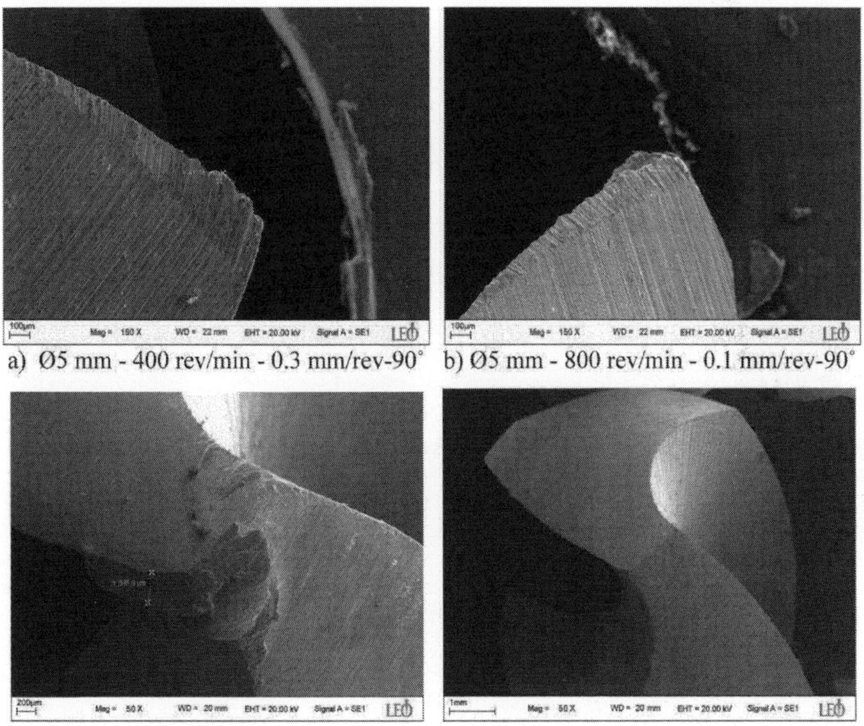

a) Ø5 mm - 400 rev/min - 0.3 mm/rev-90° b) Ø5 mm - 800 rev/min - 0.1 mm/rev-90°

c) Ø10mm - 400 rev/min - 0.2 mm/rev-90° d) Ø10 mm - 400 rev/min - 0.1 mm/rev - 90°

Figure 5. SEM images of the abrasion formed on the cutting edges

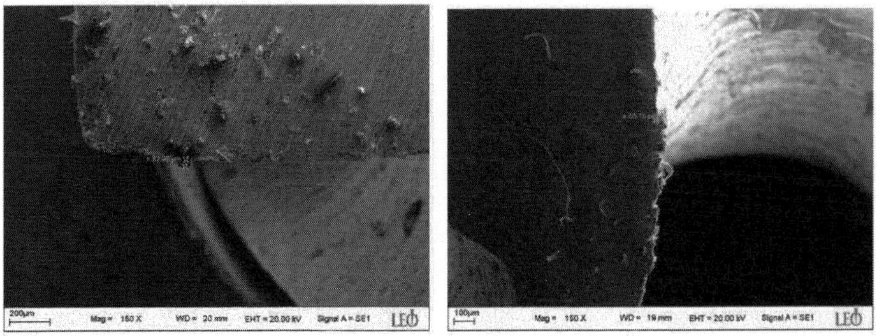

a) Ø5 mm- 400 rev/min - 0.2 mm/rev-118° b) Ø5 mm- 400 rev/min - 0.3 mm/rev-118°

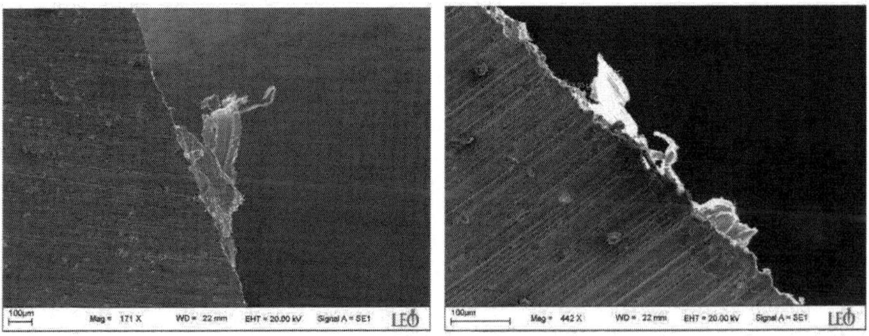

c) Ø10 mm- 1200 rev/min.- 0.3 mm/rev-118° d) Ø10 mm- 400 rev/min.- 0.2 mm/rev-130°

Figure 6. SEM images of the fractures formed on the cutting edges

In a similar manner as BUE formation, the chips adhering to helical channels obstructed the effective removal of the chips by plugging the helical channels partially (Figure 7). This case was more prominent at low rotation speeds, and the roughness increased depending on the this case (Figures 2 a-f).

As seen from the Figure 2 a-f, surface roughness increased with the increase of the feed rate, since the amount of the chips increased due to the increase in the feed rate. Because the increase of the feed rate caused high feed rate, low shear angle and thick chip formation [26]. This case signified that machined surfaces were more influenced from the forces during the cutting process. Likewise, the increase in the feed rate resulted in high friction resistance, pressure and increase in temperature [23,25]. In this case, the chip experienced to shear tensions, and adhered to the cutting tool. The amount of the chip adhesion increased depending on the feed rate, and an unstable structure formed. In order to explain this case better, chip smearing formed on the cutting edge were also investigated (Figure 8). It was supposed that notch effect of these chips on the machined due to the adhesion between the chip and cutting edge caused a corruption on the surface quality (Figure 8). This case resulted in the increase in surface roughness.

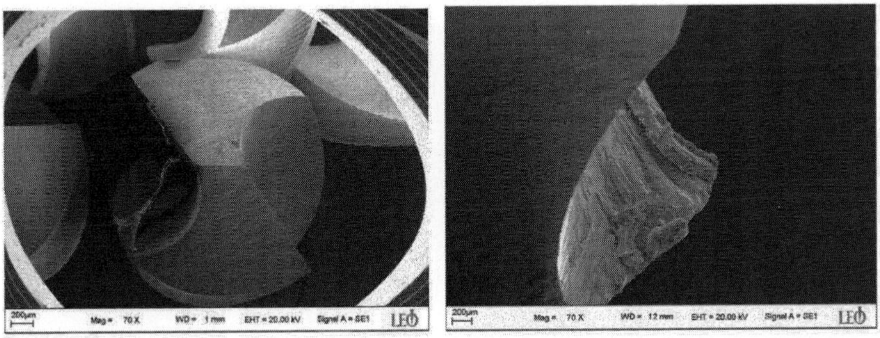

a) Ø5 mm- 400 rev/min- 0.2 mm/rev-90° b) Ø5 mm- 400 rev/min.- 0.3 mm/rev-90°

Figure 7. SEM images of the chips adhering to the helical channels.

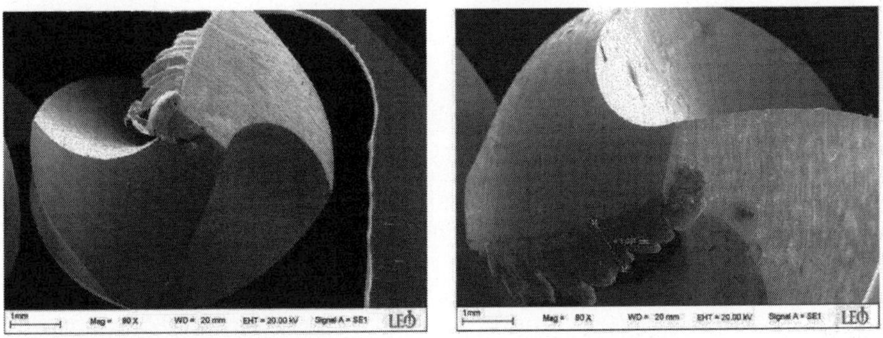

90°-0.1 mm/rev-800 rev/min 90°-0.3 mm/rev-400 rev/min

Figure 8. SEM images of the smearing formed on the cutting tool.

In addition, surface roughness changes depending on the feed rate and cutting radius in turning as denoted $Ra=0.321(f^2/r)$ in the Ref. [27]. Where R_a is the roughness, f is the feed rate, r is the radius.

Monagham and O'Reily, determined that this relation was valid for the drilling process, and a similar relation emerged in the drilling of Al 5005 (Figure 2 a-f).

Similarly, as seen from Figure 2 a-f surface roughness improved with the increase of drill point angle. This case can be explained as follows: The values of plastic deformation region, cutting edge length and chip thickness obtained for the drills having point angles of 118°, 130° and 140° are greater than those obtained for the drill having point angle of 90°.

Furthermore, since the cutting tool was worn away faster due to the expansion of the friction surface of the cutting edge with the decrease of the point angle [28] the stability of the cutting process was influenced negatively, therefore the roughness increased as seen from the Figure 2 a-f. Also, the pressure applied on the hole surface was decreased owing to the decrease in radial force with the increase of the point angle. Hence, the roughness arisen on the surface was less in comparison with the drills having small point angles.

On the other hand, it was supposed that surface roughness was also influenced by BUE arisen on the cutting edge. The change in BUE on the cutting tool depending on the different point angles was illustrated in Figure 9. As seen, as the point angle decreased, BUE influenced the form of the cutting tool more negatively. Therefore, this case affected the stability of the cutting tool during the cutting process, and caused an increase in surface roughness (Figure 2 a-f).

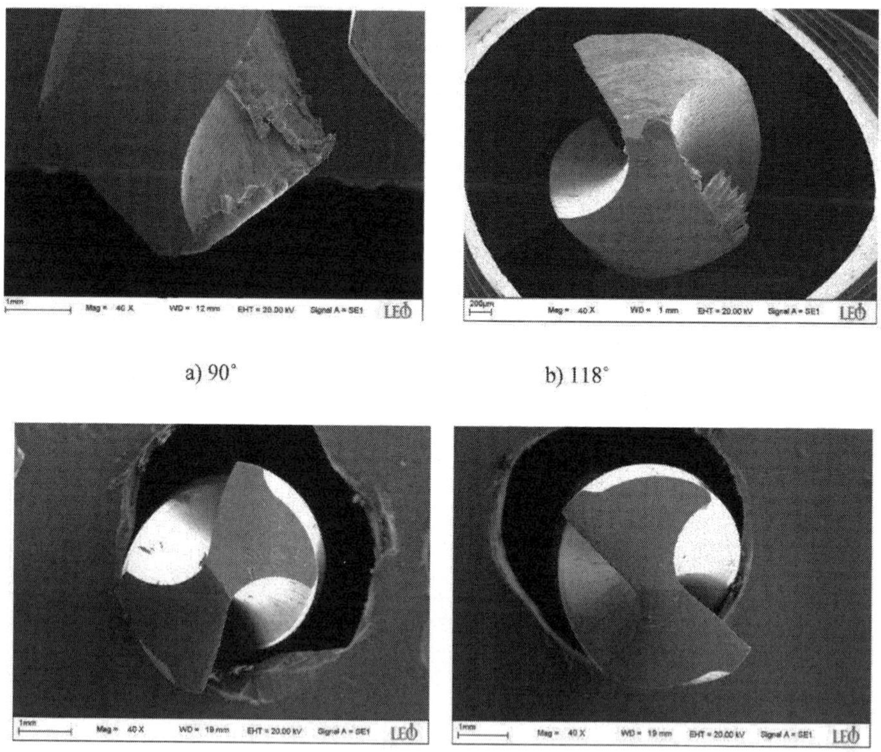

Figure 9. SEM images of BUE formation on the cutting edges at different point angles (1200 rev/min-0.1 mm/rev)

In addition, it was seen in the drilling process that the roughness values obtained for a drill with a diameter of Ø10 mm was bigger than those obtained for a drill with a diameter of Ø5 mm (Figure 2 a-f). This case was attributed to the increase of the forces due to the increase of the length of the cutting edge [23,28]. Also, as seen from Figure 10, it was supposed that the expansion of the deformation area due to the increase of drill diameter resulted in a rise in BUE formation which caused an increase in the roughness (Figure 2 a-f).

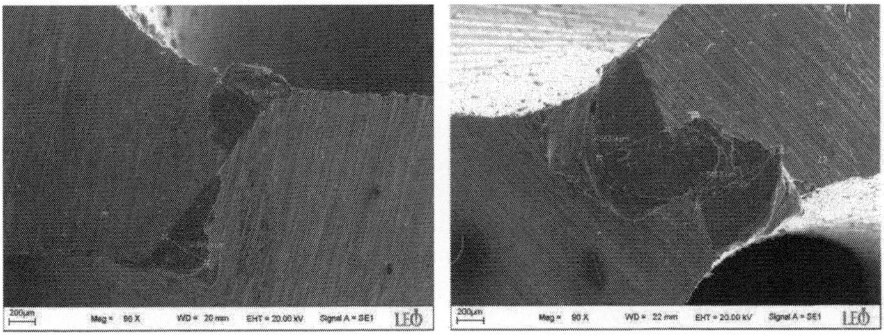

a) Ø5 mm b) Ø10 mm

Figure 10. SEM images of BUE formation on the cutting edges at different drill diameters. (800 rev/min-0.2 mm/rev-118°)

4. Conclusion

In this study, BUEs arisen on the cutting edges and the effect of drilling parameters (rotation speed, feed rate, drill diameter and point angle) on the surface roughness of the work piece were investigated experimentally in the drilling process of Al 5005 alloy on CNC milling machine. The inferences achieved were presented as follows:

i. The surface roughness decreased with the increase of the rotation speed and point angle, while it increased with the increase of the feed rate and drill diameter.
ii. It was observed that BUE arisen on the cutting edges caused fractures and wears on the cutting edges during the removal from the cutting edges.
iii. BUE formation decreased with the increase of the rotation speed, therefore the value of surface roughness decreased.
iv. BUE plugged the helical channels partially by adhering on them, and obstructed the effective removal of the chips.
v. BUE spoiled the form of the cutting tool more as the drill point angle decreased, and this case resulted in an increase in the surface roughness.

Author details

Erkan Bahçe*
İnonu University, Department of Mechanical Engineering, Malatya, Turkey

Cihan Ozel
Firat University, Department of Mechanical Engineering, Elazig, Turkey

5. References

[1] Smith, W.S., 2001. translation: Mehmet Erdoğan. Structure and Properties of Engineering Alloys. Vol.:2,

[2] Stringer, P., Byrne, G. ve Ahearne, E., 2010. ''Tool design for burr removal in drilling operations'', http://www.ucd.ie/mecheng/ams/news_items/Peter%20Stringer. pdf, 1-7.

[3] Kim, J., ve Dornfeld, D. A., 2002, ''Development of an Analytical Model for Drilling Burr Formation in Ductile Materials'', Transactions of the ASME, Vol. 124, 192-198.

[4] Puertas, I., Luis perez, C.J., 2003,'' Surface rougness prediction by factorial design of experiments in turning processes", Journal of Materials Processing Technology 143 – 144.

[5] Çoğun, C., Özses, B., 2002, "Effect Of Machining Parameters On Surface Roughness In Cnc Machine Tools", Gazi Üniv.Müh.Mim.Fak. Der. Vol. 17, No1, 59 – 75.

[6] Koelsch, J., 2001, ''Divining Edge Quality by Reading the Burrs'', Quality Magazine, December, 24-28.

[7] Karayel, D., 2008, "Prediction and control of surface roughness in CNC lathe using artificial neural network", Journal of Materials Processing Technology, 209: 3125–3137.

[8] Çetin, M. H., Özçelik, B., Kuram, E., Şimşek, B. T., Demirbeş, E., 2010,"Effect of Feed Rate on Surface Roughness And Cutting Force In The Turning of AISI 3041 Steel With Ep Added Vegetable Based Cutting Fluids", Selçuk University, II. National Machining Symposium, UTİS 2010, S:92-107, Konya, Turkey

[9] Nouari, M.; List, G.; Girot, F. & Gehin, D. 2005, Effect of machining parameters and coating on wear mechanisms in dry drilling of aluminium alloys. International Journal of Machine Tools & Manufacture, Vol. 45, No. 12-13, pp. 1436-1442, ISSN 0890-6955

[10] Lin T.R., 2002, Cutting behavior using variable feed and variable speed when drilling stainless steel with TiN-coated carbide drills. International J. Advance Manufacturing Technology. 19: 629-636.

[11] Lin, T. R. ve Shyu, R. F., 2000. ''İmprovement of Tool Life and Exit Burr using Variable Feeds when Drilling Stainless Steel with Coated Drills, Int. J. Adv. Manuf. Technol, 16, 308-313.

* Corresponding Author

[12] Kurt M., Kaynak Y. , Bakır B., Köklü U., Atakök G., Kutlu L., 2009,''Experimental Investigation And Taguchi Optımızatıon For The Effect Of Cuttıng Parameters On The Drilling Of Al 2024-T4 Alloy With Diamond Like Carbon (DLC) Coated Drills'', 5. Sixth International Advanced Technologies Symposium (IATS'09), Karabük, Turkey

[13] Dudzinski, D., Devillez, A., Moufki, A., Larrouquere, D., Zerrouki, V., Vigneau, J., 2004, ''A review of developments towards dry and high speed machining of Inconel 718 alloy'', Int. J. Mach. Tools Manufact. 44 (4), 439–456.

[14] Kılıckap, E., 2010, ''Modeling and optimization of burr height in drilling of Al-7075 using Taguchi method and response surface methodology'', Int J Adv Manuf Technol, DOI 10.1007/s00170-009-2469-x.

[15] Kılıçkap, E., 2009, ''Investigation of The Effect of Cutting Parameters on The Burr Formatıon In Drıllıng of Al-7075', I. National Machining Symposium, 142-150, Istanbul, Turkey

[16] Hanyu,H.; Kamiya,S.; Murakami,Y.; Saka,M., 2003,"Dry and semi-dry machining using finely crystallized coating cutting tools" Surface and Coatings Technology 173-174

[17] König W. and Grass P., 1989, "Quality Definition and Assessment in Drilling of Fibre Reinforced Thermosets", CIRP Annals, Vol. 38, No. 1, pp. 119-124.

[18] Tosun, N., 2006, ''Determination of optimum parameters for multi-performance characteristics, in drilling by using grey relational analysis'', *Int J Adv Manuf Technol* 28, 450–455.

[19] Sur G., Çetin H., Çevik E., Ahlatçı H., Sun Y., 2011, ''Determining the Influence of Ti Additive on Surface Roughness During Turning of AA6063 Alloy'', 6th International Advanced Technologies Symposium (IATS'11), Elazığ, Turkey.

[20] S. M. Darwish, A., M. El-Tamitni, 1997, ''Formulation of Surface Roughness Models for Machining Nickel Super Alloy with Different Tools'', Materials and Manufacturing ProcessesVol.12, No:3, 395-408,

[21] Tosun N., Kuru C. Altıntaş E. ve Erdinç E., 2010, 'Investıgatıon of Surface Roughness In Milling With Air And Conventional Cooling Method'', J. Fac. Eng. Arch. Gazi Univ. Vol 25, No 1, 141-146,.

[22] Obikawa, T., Kamata, Y., Shinozuka, J., 2006, ''High-speed grooving with applying MQL'', International Journal of Machine Tools & Manufacture, 46, 1854–1861,

[23] Şahin, Y., 2000, ''Principles of Metal Removing '', Vol. 1-2, Nobel Publisher.

[24] Kılıckap, E., Huseyinoglu M. and Ozel C., ''Emprical Study Regarding the Effects of Minimum Quantity Lubricant Utilization on Performance Characteristics in the Drilling of Al 7075'', J. Of the Braz. Soc. Of Mech. Sci.&Eng. Vol.XXXIII, No.1/53

[25] Özcelik, B., ve Bagci, E., 2006, ''Experimental and numerical studies on the determination of twist drill temperature in dry drilling'': A new approach, P I Mech Eng L-J Mat, 27, 920–927.

[26] Çakır, C., ''The Fundamentals of Modern Machining'', 1999, Uludağ University, Bursa, Turkey

[27] Monagham, J., O'Reily, P., 1992, ''The drilling of an Al/SiC metal matrix composite'', J Mater Process Tech, 33, 469-480.

[28] Akkurt, M., 2004, ''Metal Removing Methods And Machine'', Birsen Publisher, Istanbul, Turkey

Chapter 3

Effect of Abrasive Particle Size on Abrasive Wear Resistance in Otomotive Steels

Ibrahim Sevim

Additional information is available at the end of the chapter

http://dx.doi.org/10.5772/55914

1. Introduction

Machine parts are subject to the following wear types: Abrasive, adhesive, fatigue and corrosive. In abrasive wear, chipping of harder material a micrometer scale occurs as result of rubbing the soft member. The wear is formed as result of cutting, hitting, and scratching. Abrasion takes places at the solid-solid, particle-solid, solid-liquid interface [1].

If one of the surfaces which are in touch is rough and hard, it chips the other surface due to relative motion or touching forces. The wear is called two-body abrasive wear. If there are free abrasive particles between the two bodies, the wear is called three-body abrasive wear. The free abrasive particles may be external material dust or the remains of chipping. Usually, the wear starts as a two-body abrasive or adhesive wear and then becomes a three-body wear as dust form between the two surfaces due to external particles, chipping remains, or oxide particles. In three-body abrasive wear, wear rate increases as diameter of abrasive particles increases. Gouging, high stress abrasion and low stress abrasion are types of three-body abrasive wear [1-3].

In gouging, surface wear is formed using large abrasive particles. A gouging mechanism is common in ground leveling machines and excavation and digging machines. In these machines, wear occurs on moving, digging, and excavating members. High-stress abrasion occurs when the sharp edged small abrasive particles, which are formed by the crushed particles under excessive loads, scratch the surface. Ball-bearing grinders are mostly subject to high-stress abrasion [2]. These grinders are predominately used to crush the metallic ores and minerals. High-stress abrasion contributes to the significant portion of the wear in grinders. Low-stress abrasion takes place when there is no crushing or grinding in the abrasive particles and one of the surfaces is subject to wear [3].

Abrasive wear experiments have been made with substances containing one or more abrasive. Abrasive statements, which are obtained through single abrasive end patterns (i.e.

sphere, pyramid, and cone) are adapted to abrasive wear cases with abrasive particle more than one based on some assumptions. The abrasive particle is generally modeled as a cone shape [4]. Rabinowicz [5] has aimed at a simple expression for the volume of material removed during two-body abrasion by a conical abrasive particle;

$$\frac{V}{L} = \left(\frac{2\tan\alpha}{\pi}\right)\left(\frac{F_N}{H}\right) \tag{1}$$

where V is the volume loss due to wear, L the sliding distance, F_N the normal load on the conical particle and H hardness of wearing surface and α the attack angle of the abrasive particle.

The equation (1), for linear wear density can be written as follows [1];

$$W = k\frac{P}{H} \tag{2}$$

Where; W: linear wear density, k: wear coefficient, P: pressure applied on surface, H: hardness of abraded material.

For pure metals and annealed steels, the wear resistance versus hardness is a line passing through the origin. The linear zone is called zone *I* throughout the paper. The abrasive wear resistance versus hardness graph of the heat-treated steels is a line not passing through the origin [3]. This behavior cannot be derived from (2). The zone corresponding this is called zone *II* . The zones *II* and *I* are shown in Figure 1 [1, 4]. (2) is similar to the Archard's adhesive wear expression.

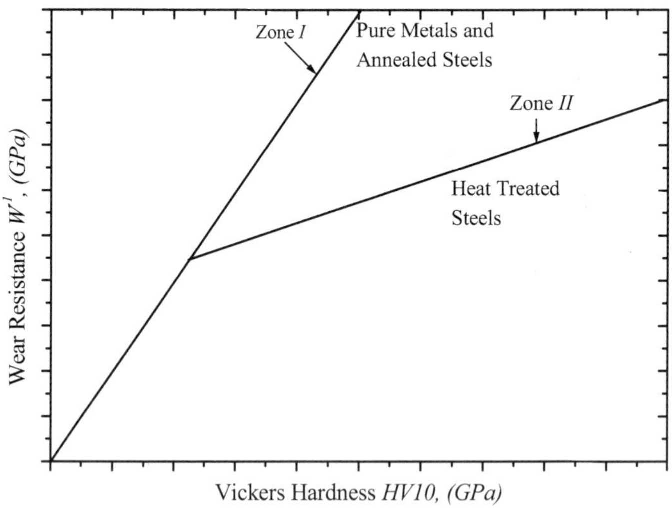

Figure 1. Relationship between wear resistance and hardness [1, 4].

Generally, (2) does not agree with the experimental results. The main reasons for this incompatibility are the changes of wear coefficient k depending on abrasive grit size [5, 6]. In literature, there are many investigations about the effect of the abrasive grit size on abrasive wear rate in zone *I*. Avient et al [7] have examined the abrasive behavior of many materials and realized that the clogging of the interstices between the finer abrasive grains by wear debris is responsible for the grit size effect. This decreases the number of abrasive grains, which contact the surface and remove material, thus decreasing the abrasive wear rate. Mulhearn and Samuel [8] studied samples of silicon carbide (SiC) abrasive papers. They believe that the mechanical properties of coarse and of fine abrasive grains are different, and that the fine grains have a needle-like shape and contain many cracks, thus braking up more readily. In this way, abrasive wear rate becomes zero, because small grains are flattened. Rabinowich and Mutis [9], have aimed an account of the size effect using adhesive wear particles. Using a surface energy criterion, they theoretically show that the critical abrasive particle size is a function of the adhesive particle size of the material being worn away. Sin et.al [5] have used the critical depth of penetration to explain the effect of grit size on abrasive wear loss and have found that there was not a critical abrasive particle size for a specific material. They also showed that the constant wear rate starts at 80 μm abrasive particle size for all metals used in the experiments. The elastic contact hypothesis was first suggested by Larsen-Badse [10] who measured the size and number of grooves formed on polished copper specimens abraded by SiC abrasive papers and estimated the real contact area. He postulated that many fine grit have elastic interaction with the surface. It was also suggested that the fraction of the load carried by particle in elastic contact increased with decreasing grit size since it is unlikely that the abrasive grits gradually become more angular with increased size. Moore and Douthwaite [11], have tried to explain the size effect by plastic deformation concept below worn surfaces. They estimated the equivalent plastic strain and the flow stress as a function of depth below worn surface and calculated the work done in deforming the material below the groove and energy absorbed in plowing the surface. They concluded that the energy expended in plastic deformation of material to form the grooves and deform the surface account for almost all the external work done for all grit sizes in abrasion and that wear volume is dependent on the grit size probably because the deterioration and pick up of abrasive particles become more intense at small grit sizes. Hutchings [12], has stated that the size effect is due to the variation of shape changing rate dependent to abrasive particle size. However, Misra and Finnie [13], have found that the shape-changing rate has only changed the wear resistance, and has no effect on the dependency of abrasive particle size. Many researchers have examined the abrasive particle size effect in the zone *II* [14, 15]. Rabinowich [14] determined empirically the following abrasive wear rate expression for the *zone II* using only one type of abrasive particle size

$$W = k\frac{P}{\frac{1}{3}H + \frac{2}{3}H_O} \tag{3}$$

where H is the hardness of alloy, H_O is the hardness of the alloy in the fully soft condition and P is the pressure applied to the surface.

Khruschov [15] has studied experimentally the zone *I* in a stationary abrasive particle size using the non-heat treated steels and he found the relative wear resistance –hardness relationship for metals as follows;

$$\varepsilon = bH \tag{4}$$

where ε is the relative wear resistance, b a constant coefficient and H the initial hardness.

Furthermore the following relationship has been determined to be in zone *II*, between the relative wear resistances of heat-treated steels and hardness;

$$\varepsilon = \left(\varepsilon_0 - C_0 H_0\right) + C_1 H \tag{5}$$

where ε_0 and H_0 are the relative wear resistance and hardness of annealed steel, and C_0 and C_1 are constants.

There are numerous explanations in the literature to explain the abrasive grit size effect. However, most of them have been insufficient since they have not been able to explain the grit size effect encountered in all abrasive wear mechanisms (for example erosive wear) [15-18].

The focus of this study is to investigate the effect of abrasive particle size on abrasive wear resistance in zone *I, II* and to develop the equations of empirical abrasive wear resistance connected to abrasive particle size. Moreover, to search for the effects of relative wear resistance in zone *I, II* and to develop the equations of empirical relative wear resistance connected to abrasive particle size.

2. Experimental procedure

The steels AISI 1010, 1030, 1040, 1050 and 50CrV4 were used in the study. The chemical compositions of these samples are given in Table 1. The specimens were in the form of cylinders of 9 *mm* diameter and 3 *mm* height.

The samples were prepared from non-heat treated and heat-treated steels. The heat treatment conditions are given in Table 2. The samples were ground with abrasive papers grading from 80 to 800 meshes and then polished with 0.3 μm diamonds. The hardness were measured by the Vickers hardness method in load of 98.0865 N (*HV*10). The average of measurements and the standard deviations were calculated. The average hardness values and standard derivations are given in Table 2. Wear experiment was carried out on the pin-abrasion testing machine shown in Figure 2; tambour diameter D=118 *mm*, tambour rotation n=1000 *rpm* and abrasive wear set-up rate V=6.18 ms^{-1}. In wear experiments, the 180, 125, 85, 70 and 50 μm alumina (Al_2O_3) abrasive paper in sizes 100x1150 *mm* were used. For wear experiments, the apparatus in Figure 3 was mounted on the pin-abrasion testing machine. In order to fix the samples within apparatus in Figure 3, the cylindrical copper bars of 50 *mm* in length and 20 *mm* in diameter have been used. In order to prepare the specimens for abrasive wear test, holes of 9 *mm* in diameter and 1.5 *mm* in depth were milled on one end of the copper bars through which the specimen were replaced.

Alloys	C (%)	Si (%)	Mn (%)	P (%)	S (%)	Cr (%)	Mo (%)	Ni (%)	Al (%)	Cu (%)	Ti (%)	V (%)
1010	0.107	0.11	0.413	0.019	0.025	_	0.003	_	0.032	0.031	0.002	_
1030	0.328	0.069	0.673	0.015	0.019	_	0.001	_	_	0.037	0.002	0.005
1040	0.402	0.247	0.82	0.012	0.028	0.025	0.001	0.003	0.014	0.032	0.001	0.003
1050	0.506	0.252	0.654	0.014	0.006	0.251	0.002	_	0.006	0.017	0.002	0.006
50CrV4	0.523	0.394	0.915	0.021	0.027	0.917	0.025	0.034	_	0.183	_	0.095

Table 1. The chemical compositions of experiment sample (wt. %) [19, 20]

Materials	Heat Treatment	Vickers hardness *HV10 (MPa)*
AISI1010	-	1648±10
AISI1030	-	1716±20
AISI1040	-	1961±29
AISI1050	-	2175±34
50CrV4	-	2549±49
AISI1010	Water quenched from 900-925 °C	2255±54
AISI1030	Water quenched from 830-850 °C	5609±20
AISI1040	Water quenched from 820-850 °C	6276±15
AISI1050	Water quenched from 810-840 °C	6570±0
50CrV4	Water quenched from 830-850 °C	8895±0
AISI1010	Water quenched from 900-925 °C + 2 hours refrigerated at –25 °C	2256±10
AISI1030	Water quenched from 830-850 °C + 2 hours refrigerated at –25 °C	6767±25
AISI1040	Water quenched from 820-850 °C + 2 hours refrigerated at –25 °C	7100±39
AISI1050	Water quenched from 810-840 °C + 2 hours refrigerated at –25 °C	7875±20
50CrV4	Water quenched from 830-850 °C + 2 hours refrigerated at –25 °C	8895±0
AISI1010	Water quenched from 900-925 °C + tempered at 250 °C	1873±25
AISI1030	Water quenched from 830-850 °C + tempered at 250 °C	5551±34
AISI1040	Water quenched from 820-850 °C + tempered at 250 °C	5943±17

Materials	Heat Treatment	Vickers hardness *HV10 (MPa)*
AISI1050	Water quenched from 810-840 °C + tempered at 250 °C	6139±37
50CrV4	Water quenched from 830-850 °C + tempered at 250 °C	6845±25
AISI1030	Water quenched from 830-850 °C + tempered at 350 °C	4511±83
AISI1040	Water quenched from 820-850 °C + tempered at 350 °C	4884±26
AISI1050	Water quenched from 810-840 °C + tempered at 350 °C	5198±49
50CrV4	Water quenched from 830-850 °C + tempered at 350 °C	5492±29
AISI1030	Water quenched from 830-850 °C + tempered at 450 °C	3118±26
AISI1040	Water quenched from 820-850 °C + tempered at 450 °C	4550±49
AISI1050	Water quenched from 810-840 °C + tempered at 450 °C	4737±20
50CrV4	Water quenched from 830-850 °C + tempered at 450 °C	4805±39
AISI1030	Water quenched from 830-850 °C + tempered at 550 °C	3030±55
AISI1040	Water quenched from 820-850 °C + tempered at 550 °C	3324±29
AISI1050	Water quenched from 810-840 °C + tempered at 550 °C	3589±35
50CrV4	Water quenched from 830-850 °C + tempered at 550 °C	3727±64
AISI1030	Water quenched from 830-850 °C + tempered at 650 °C	1973±10
AISI1040	Water quenched from 820-850 °C + tempered at 650 °C	2059±25
AISI1050	Water quenched from 810-840 °C + tempered at 650 °C	2256±39
50CrV4	Water quenched from 830-850 °C + tempered at 650 °C	2902±34

Table 2. Heat treatment and hardness values [19, 20]

Figure 2. The pin-abrasion testing machine [19, 20]

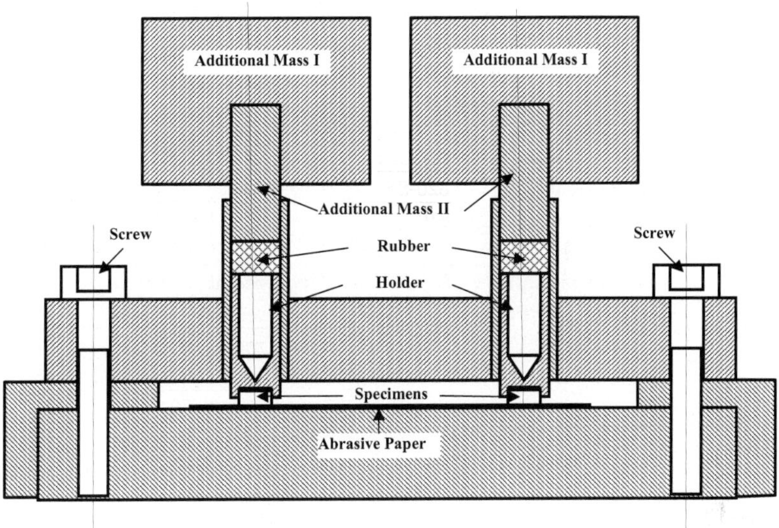

Figure 3. Apparatus for abrasive wear experiments [19, 20]

On the other end, a hole of 14 *mm* in diameter and 25 *mm* in depth was drilled in order to balance the sample. An adhesive were applied to the samples and then the samples were attached into the holes milled on copper bars. Prior to the experiment, the samples were cleaned with alcohol and the mass of the sample were measured gravimetrically with 10^{-4} *mg* sensitivity. Then, they were assembled into the apparatus (Figure 3) mounted on the pin-abrasion testing machine. Hard rubber dampers of 20 *mm* diameter and 10 *mm* thickness were put on the experiment sample to damp out the vibrations. Additional masses were fixed on the copper bars that were on top the rubber dampers. Abrasive wear experiments have been performed on each sample for 10 seconds under 0.13 *MPa* pressure and the experiment were repeated 5 times under the same conditions on each sample. At each repetition, the mass of the sample were determined gravimetrically and recorded. The wear volumes, V, were determined from the measured mass losses using the specific mass of the samples. The linear wear rates, W, were computed using the following equation

$$W = \frac{V}{LA} \tag{6}$$

where L is the sliding distance of the experiment sample and A is the wear surface area of the sample.

3. Results and discussion

In this section, we use the abrasive wear expressions and definitions given in nomenclature section. In particular, we define the pressure wear resistance, W_p^{-1} as:

$$W_P^{-1} = \frac{P}{W} \tag{7}$$

where P is the applied pressure to the experiment sample, and W is the linear wear rate defined in (7).

3.1. For non-heat treated steels

The relationship between the pressure wear resistance, W_P^{-1}, and hardness, H, of non-heat treated steels is illustrated in Figure 4. The following relationship can be deducted via curve fitting using the least square method in Figure 4;

$$W_P^{-1} = C_2 H \tag{8}$$

where $C_2 = k^{-1}$, and k is the wear coefficient.

Rewriting (8) in terms of wear coefficient, the following expression for pressure wear resistance is obtained

$$W_P^{-1} = \frac{H}{k} \tag{9}$$

In Table 3, the coefficients C_2, k and R are given for non-heat-treated steels. The variation of wear coefficients k (Table 3) with abrasive particle size d for non-heat treated steels is seen in Figure 5.

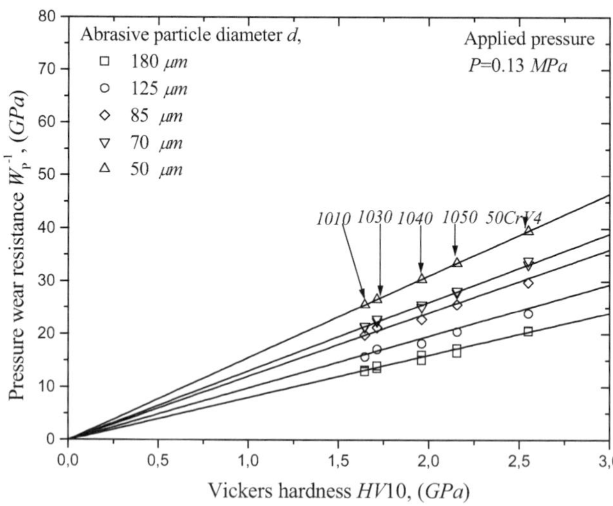

Figure 4. Non-heat-treated steels pressure wear resistance versus Vickers hardness (Parameter: Abrasive particle size) [19]

Materials	Abrasive particle size d, (μm)	C_2	Wear coefficient $k=1/C_2$	Coefficient of Correlation R
Non-heat treated steels	180	8	0.125	0.99
	125	9.8	0.111	0.99
	85	12	0.083	0.99
	70	13	0.077	0.99
	50	15.5	0.065	1

Table 3. Coefficient C_2 and wear coefficient k[19]

As seen in Figure 5, the dependence of wear coefficient k on the abrasive particle size d is consistent with previous works [5, 6, 10, 15, 16]. However, the results in Figure 5 shows that although wear coefficient k increases initially fast with increasing abrasive particle size d, the wear coefficient does not reach to a steady state value in terms of a critical particle size. Besides, as long as the abrasive particle size increases, the slope of the curve decreases as seen Figure 5.

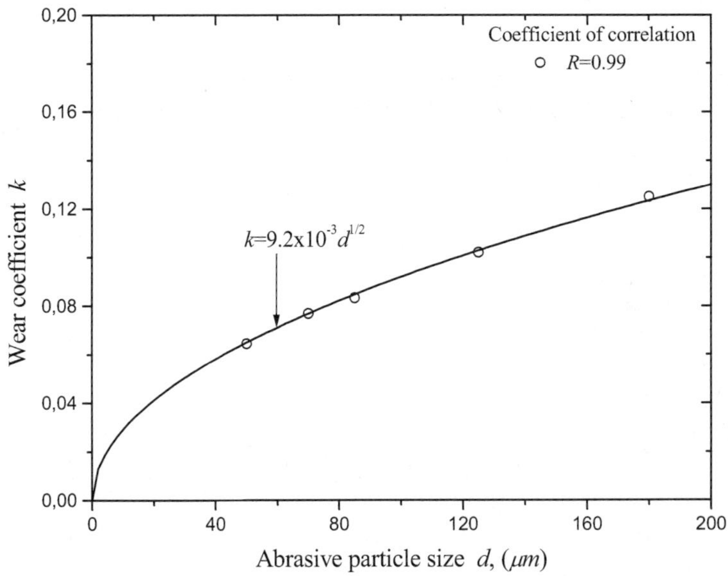

Figure 5. Variations of wear coefficient k of non-heat treated steels versus abrasive particle size [19]

From Figure 5, the relation between wear coefficient k and particle size d for zone I is given by

$$k = 9.2\sqrt{d} \tag{10}$$

where d is abrasive particle size.

If (10) is substituted in (9), the pressure wear resistance expression for zone I becomes

$$\left(W_P^{-1}\right)_{ZoneI} = \frac{H}{9.2\sqrt{d}} \tag{11}$$

and the wear resistance is

$$\left(W^{-1}\right)_{ZoneI} = \frac{1}{9.2\sqrt{d}}\frac{H}{P} \tag{12}$$

The previous works [3, 5, 6] states that the wear coefficient k and/or the wear rate W are dependent on the particle size d for pure metals and non-heat treated steels, but they did not give the mathematical expressions for this. In this study situation, the equation (12) was derived for the relation between the wear coefficient k and the particle size d using a curve fitting technique based on least square approximation for non-heat treated steels. (12) is valid for ideal microcutting, according to Zum Garh [3].

3.2. For heat treated steels

The variation of pressure wear resistance of the heat-treated steels (water quenched, water quenched+ refrigerated at –25 °C, water quenched + tempered) with hardness is given in Figure 6. According to Figure 6, the general expression of pressure wear resistance in terms of hardness for heat-treated steels can be written as follows;

$$\left(W_P^{-1}\right)_{ZoneII} = C_3 + C_4 \tag{13}$$

where C_3 and C_4 are constants.

C_3 and C_4 constants and coefficient of correlation R are given in Table 4 for heat-treated steels. (3) shows how the pressure wear resistance in zone II changes with the hardness. Let us define C_3 and C_4 as follows:

$$C_3 = \frac{2}{3k}H_0 \tag{14}$$

$$C_4 = \frac{1}{3k} \tag{15}$$

where H_0 is defined in (3) as the hardness of annealed alloyed steel.

If we substitute for C_3 and C_4 in (13), we obtain (3). According to (3), since the values of H and H_0 are dependent on abrasive particle size d, both coefficients in (13) are dependent on abrasive particle size d. But our results (Table 4) show that C_4 coefficient is not dependent on abrasive particle size d. The variation of C_3 coefficient is plotted versus abrasive particle size d (Figure 7). Since C_4 coefficient is not dependent on abrasive particle size it is understood that the abrasive particle size for heat-treated steels does not change the slope in zone *II* (Figure 6). The abrasive particle size affects the slopes in zone *I* and *II* (Figure 8). If C_3 coefficient in (13) replaced with the value from Figure 7 and C_4 coefficient from Table 4, the pressure wear resistance expression in zone *II* for the heat-treated steels becomes;

$$\left(W_P^{-1}\right)_{ZoneII} = \frac{1.4}{d} + 2.6H \tag{16}$$

and the wear resistance becomes

$$\left(W^{-1}\right)_{ZoneII} = \frac{1}{P}\left(\frac{1.4}{d} + 2.6H\right) \tag{17}$$

Materials	Abrasive particle size d, (μm)	C_3	C_4	Coefficient of Correlation R
Heat treated steels	180	7750	2.6	0.98
	125	11700	2.6	0.98
	85	16600	2.6	0.97
	70	18200	2.6	0.96
	50	28800	2.6	0.99

Table 4. Coefficients C3 and C4 [20]

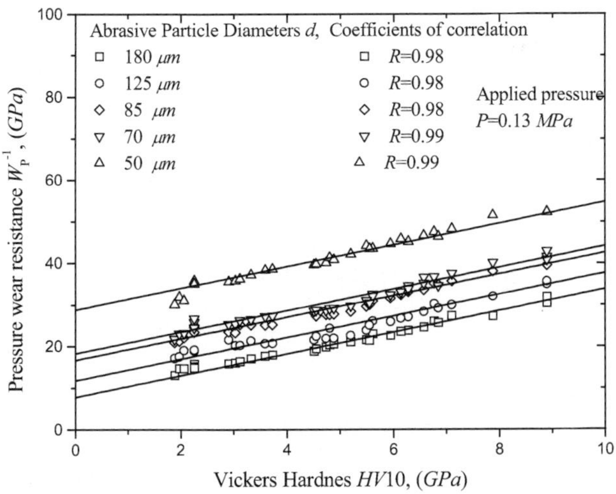

Figure 6. Heat-treated steels pressure wear resistance versus Vickers hardness (Parameter: Abrasive particle size) [20]

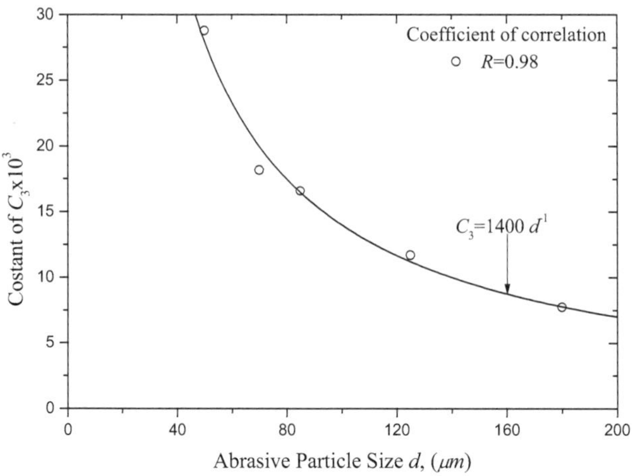

Figure 7. Constant C_3 of heat-treated steels versus abrasive particle size *d* [20]

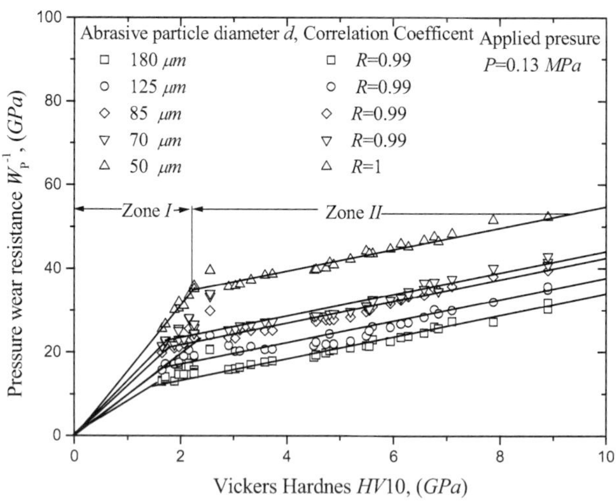

Figure 8. Non-heat treated and heat treated steels pressure wear resistance versus Vickers hardness (Parameter: Abrasive particle size) [19, 20]

In previous works, the abrasive wear resistances of heat-treated steels were found to be different than that of non-heat treated steels. Researchers concluded that this difference was due to the heat-treatment of the material [14, 15, 16]. After heat-treatment the hardness of the material changes. According to the abrasive wear mechanism of heat-treated steels, abrasive particle cuts more chips than wear groove volume [1-3]. Abrasive particle produces chip via micro cutting and micro cracking mechanisms. It was concluded that the difference in the wear resistance of heat-treated and non-heat treated steels arises from micro cracking mechanism in heat-treated steels during abrasive wear.

The variations of the pressure wear resistances of non-heat-treated and heat-treated steels with the hardness are shown in Figure 8. As seen in Figure 8, the pressure wear resistances of non-heat-treated and heat-treated steels are dependent on abrasive particle size.

3.3. Relative wear resistance for non heat treated

From Figure 9, the dependence of the relative wear resistance on hardness for non-heat treated steels can be expressed as

$$e = 6x10^{-4} H \tag{18}$$

The relative wear resistance of non-heat-treated steels does not depend on abrasive particle size. This result is supported with the results calculated by equation (4) which was proposed by Khruschov [15].

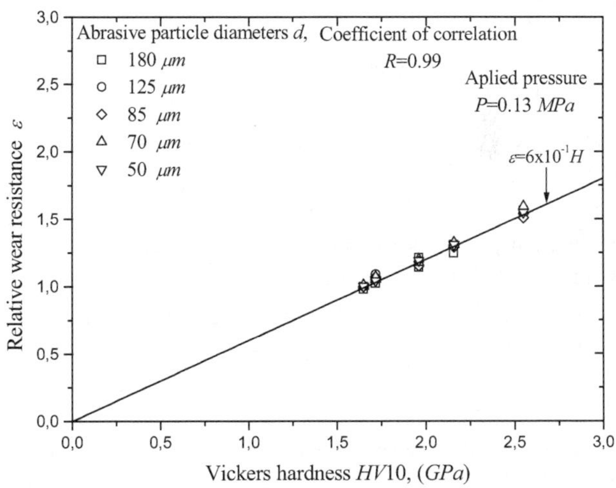

Figure 9. Non-heat-treated steels relative rear resistance versus Vickers hardness (Parameter: Abrasive particle size) [19]

3.4. Relative wear resistance for heat treated steels

The variation of pressure wear resistance of the heat-treated steels (water quenched, water quenched+ refrigerated at –25 °C, water quenched + tempered) with hardness has been shown in Figure 10. As seen in Figure 10, the relative wear resistance in steel shows different slopes depending on abrasive particle size. The relative wear resistance equations in zone *II* for the heat-treated steels can be written in general as follows;

$$\varepsilon = A_o + B_o H \tag{19}$$

where A_0 and B_0 are constant coefficients.

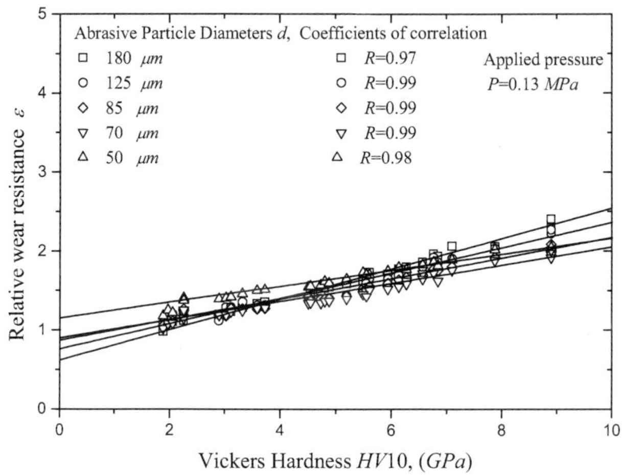

Figure 10. Heat-treated steels relative wear resistance versus Vickers hardness (Parameter: Abrasive particle size) [19, 20]

The experimental results for A_0 and B_0 constants, and coefficient of correlation R are given in Table 5 for heat-treated steels. (5) shows how the relative wear resistance in zone *II* changes with the hardness. Let us define A_0 and B_0 as follows;

$$A_0 = \left(\varepsilon_o - C_o H_O\right) \tag{20}$$

$$B_0 = C_1 \tag{21}$$

The variation of A_0 and B_0 constants are plotted versus abrasive particle size d (Figure 11). The following equation are obtained using least square approximation method,

$$A_0 = \frac{8x10^{-3}}{\sqrt{d}} \tag{22}$$

$$B_0 = 1.42x10^{-3}\sqrt{d} \tag{23}$$

Materials	Abrasive particle size d, (μm)	A_0	B_0 10^{-5}
Heat treated steels	180	0.62	19.2
	125	0.76	16
	85	0.87	13
	70	0.9	11.5
	50	1.158	10.1

Table 5. Coefficients A_0 and B_0 [20]

If A_0 and B_0 constants in (19) replaced with the expressions given in (22) and (23), the relative wear resistance expression in zone *II* for the heat-treated steels becomes;

$$\varepsilon = \frac{8x10^{-3}}{\sqrt{d}} + 1.42x10^{-3}\sqrt{d}H \tag{24}$$

The hardness H, of abraded material versus the relative wear resistances ε, of the non-heat-treated and heat-treated steels are shown graphically in Figure 12. As seen in Figure 12 and (18), the relative wear resistance ε, is independent on abrasive particle size d in zone *I* while it is dependent on d in zone *II* (see (24)).

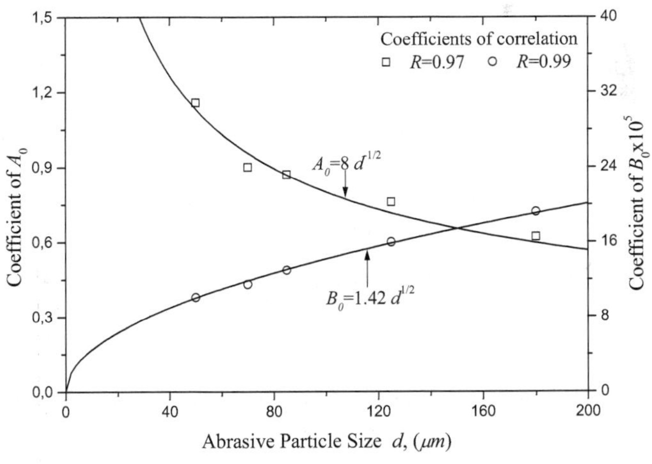

Figure 11. Coefficients A_0 and B_0 of heat-treated steels versus abrasive particle size d [19, 20]

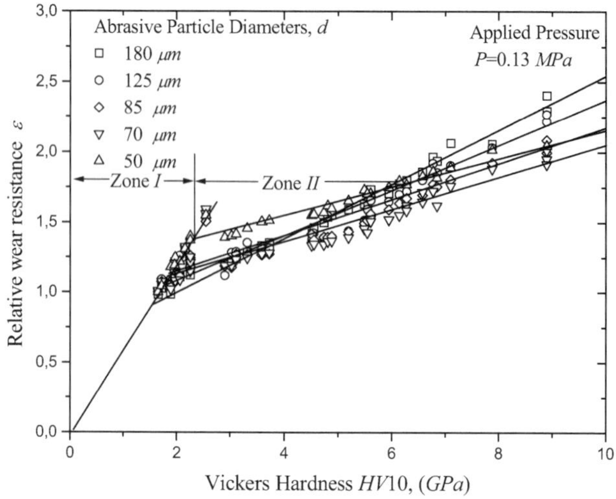

Figure 12. Non-heat treated and heat treated steels relative wear resistance versus Vickers hardness (Parameter: Abrasive particle size) [19, 20]

4. Conclusion

The results showed that the wear resistance of non-heat treated and heat-treated steels are functions of the abrasive particle size. From the results, an empirical mathematical wear resistance model and an empirical mathematical relative wear resistance ε, as a function of abrasive particle size d were derived [18-20].

- There is a linear relationship between the abrasive wear resistance W^{-1} and hardness H, depending on abrasive particle size d, for non-heat treated steels. The relationship between wear coefficient k and abrasive particle size d is a parabolic as seen in equation (10). The wear resistance W^{-1} is inversely proportional with the square root of particle size d, for non-heat treated steels as seen in equation (12).
- The relationships for the heat-treated steels between the abrasive wear resistance and hardness H, show positive intercepts on the ordinate, depending on abrasive particle size d (equation (17)).
- The relative wear resistance ε and hardness H related linearly for non-heat treated steels as it can be seen in equation (18), abrasive particle size does not effect the relationship between hardness H and relative wear resistance ε. But, relative wear resistance ε, for the heat-treated steels is dependent on abrasive particle size d, and the relationships for the heat-treated steels show positive intercepts on the ordinate. The proportionality behavior of hardness H and relative wear resistance ε, is dependent on the terms of $\frac{1}{\sqrt{d}}$ and $\sqrt{d}$ as given in equation (24).

- Heat-treated steels have lower resistance to wear than non heat-treated steels of the same hardness.

Nomenclature

$W = \frac{V}{LA} = \frac{G}{\rho LA}$: Wear Rate (Linear wear intensity)

$W^{-1} = \frac{\rho LA}{G}$: Wear Resistance

$W_p^{-1} = P\frac{\rho LA}{G}$: Pressure Wear Resistance (MPa)

$\varepsilon = \frac{W_n^{-1}}{W_r^{-1}}$: Relative Wear Resistance

W_n^{-1} : Wear Resistance of sample

W_r^{-1} : Wear Resistance of reference material

Author details

Ibrahim Sevim
Mersin University, Engineering Faculty, Department of Mechanical Engineering, Ciftlikkoy, Mersin, Turkey

5. References

[1] Sevim. I., "Effect of Abrasive Particle Size on Wear Resistance for Abrasive Wear of Steels". Ph.D. Thesis, İ.T.Ü. Institute of Science and Technology, İstanbul, 1998
[2] Rabinowicz E., A., "Friction and wear of materials" Wiley, New York, pp: 168, 1965
[3] Zum Gahr.K.H." Microstructure and Wear of Materials". Elsevier Science, 1987
[4] Hakkirigawa. K., and Li, Z., Z., The effect of hardness on the transition of abrasive wear mechanism steels, Wear of Material pp: 585-593, 1987
[5] Sin, H., Saka, N., and Suh, P., Abrasive Wear Mechanisms and The Grit Size Effect of Metals, Wear, 55(1979) 163-190.
[6] Misra, M., and Finnie, I., Some observations on two-body abrasive wear, Wear, 68(1981) pp:41-56.
[7] Avient., W., E., Goddard, J., and Wilman, H., An Experimental Study of Friction and wear during Abrasion of Metals, Proc. R. Soc. (London), Ser. A, 256 (1960) 159-179.
[8] Mulhearn, T., O., and Samuels, L., E., The Abrasion of Metals: A Model of The Process. Wear 5(1962) 478-498

[9] Rabinowicz, E. and Mutis, A., Effect of Abrasive Particle Size on Wear. Wear, 8(1965) 381-390

[10] Larsen-Badse, J., Influence of Grit Size and Specimen Size on Wear during Sliding Abrasion. Wear 12(1968) 35-53

[11] Moore, M. A. and Douthwaite, R. M. Plastic deformation below worn surface, Metall. Trans., 7A(1976) 1833-1839

[12] Hutchings, I. M., Tribology: "Friction and Wear of Engineering Materials", Edward Arnold, London. 1992.

[13] Misra, M., and Finnie, I., On The Size Effect in Abrasive and Erosive Wear, Wear, 65(1981) 359-373.

[14] Rabinowicz, E., "Penetration Hardness and Toughness Indicators of Wear Resistance", Int. Conference, Tribology-Friction, Lubrication and Wear, Volume 1, pp: 197-204, 1987.

[15] Khruschov, M., M., Principles of Abrasive Wear, Wear, 28(1974) Pp: 69-88.

[16] Misra, A., Finnie, I., A Review of the Abrasive Wear of Metals, Transactions of the ASME, Vol: 104 91-101, 1982.

[17] Sevim I., Dry sliding wear of 332.0 unaged Al-Si alloys at elevated temperatures, Kovove Materialy-Metallic Materials 44(2006), pp:151-159

[18] Sevim I., Eryurek B. Effect of fracture toughness on abrasive wear resistance of steels Materials and Design 27 (2006) pp:911–919

[19] Sevim I., Eryurek B. Effect of abrasive particle size on wear resistance in non-heat-treated steels, Kovove Materialy-Metallic Materials 43(2005), pp:158-168

[20] Sevim I., Eryurek B. Effect of abrasive particle size on wear resistance in steels Materials and Design 27 (2006) pp: 173–181

Chapter 4

A New Attempt to Better Understand Arrhenius Equation and Its Activation Energy

Andrzej Kulczycki and Czesław Kajdas

Additional information is available at the end of the chapter

http://dx.doi.org/10.5772/54503

1. Introduction

Activation energy (Ea) is strictly combined with kinetics of chemical reactions. The relationship is described by Arrhenius equation

$$k = A\exp(-E_a/RT) \tag{1}$$

where k is the rate coefficient, A is a constant, R is the universal gas constant, and T is the temperature (in Kelvin); R has the value of 8.314×10^{-3} kJ $mol^{-1}K^{-1}$, E_a is the amount of energy required to ensure that a reaction happens. Common sense is that at higher temperatures, the probability of two molecules colliding is higher. Accordingly, a given reaction rate is higher and the reaction proceeds faster and, the effect of temperature on reaction rates is calculated using the Arrhenius equation. Further reaction rate enhancement is promoted by catalysis. Catalysis is the phenomenon of a catalyst in action. Catalyst is a material that increases the rate of chemical reaction, and for equilibrium reactions it increases the rate at which a chemical system approaches equilibrium, without being consumed in the process [1]. It is applied in small amounts relative to the reactants. Chemical kinetics, based on Arrhenius equation assumes that catalysts lower the activation energy (E_a) and the presence of catalyst results higher reaction rate at the same temperature. In chemical kinetics E_a is the height of the potential barrier separating the products and reactants. Catalytic reactions have a lower E_a than those of the thermally activated. This fact enables a chemical reaction not only to proceed faster but also at a lower temperature than otherwise possible. The solid heterogeneous catalyst mechanism that would lower activation energy is still under discussion. It is known that the effect of catalysts is intrinsically connected to the material surface states. However, the connection of catalysts material states to their action is not yet fully clear and specific stimulators of this action are unknown. Few years ago [2] a new approach to E_a was proposed and a first indirect confirmation was made [3]. Dante et al. [4]

supported the new approach to E_a by theoretical considerations based on irreversible thermodynamics.

Heterogeneous catalysis provides the link between reactants and products on a reaction pathway which involves simultaneous motion of several to very many atoms [5]. Predictability of the outcome of catalytic reactions is controlled by their molecular mechanisms. Thus, the importance of the activation energy better understanding can not be overestimated. Some forty years ago, work [6] demonstrated that a "chemically stimulated" exo-electron emission (EEE) occurs simultaneously during the partial oxidation process of ethylene. It was also found that the emission rate was proportional to the rate of ethylene oxide formation. Therefore, discussing heterogeneous catalytic reactions, EEE process should also be taken into account, because they involve electro physical phenomena. Additionally, such electrons are of low-energy and are produced from the excited active catalytic surfaces.

2. Background and the present work goal

2.1. Recent papers focused on the activation energy (E_a) new approach

Paper [2] has proposed the new hypothesis concerning kinetics of chemical reactions and Ea. Shortly the hypothesis was confirmed indirectly [3] by using earlier published results on energy angular distribution of electrons emitted from MIM systems [7] and angular distribution of photoelectrons emitted from solids [8]. The next confirmation step was based on irreversible thermodynamics [4] and, it was demonstrated that the equation of the found reaction rate is also consistent with two different pathways for tribochemical reactions:

$$J_c = \{\nu_i \exp(-E_a/RT)\,(1 - e^{-A/RT}) + k\,[1 - \exp(-J_u c t \tau/T)]\,\}\,\Pi a_i^{\nu_i} \tag{2}$$

where: J_c – reaction rate, J_u – energy flow towards the reagents due to triboreactions, τ – shear stress due to friction, A – chemical affinity, c and k – constants, t – time, a_i – activities of reagents, ν_i – stoichiometric coefficient.

The first one is the thermal mechanism, typical for non-friction conditions and, the second concerns the direct transfer of energy from triboelectrons to molecules. The latter one generates special excited or activated molecules, such as radicals or ions which react rapidly to form the products, enhancing the global reaction rate.

2.2. Aim of the paper

The major goal of the present paper is to enhance the developed idea and provide it with the better understanding of Arrehnius equation and its activation energy. To make a progress in that field, α_i model developed by Kulczycki [9] was here applied along with its detailed thermodynamic interpretation. Detailed disscussion is focused on the basic Arrehnius equation.

3. Thermodynamic interpretation of α_i model

3.1. Theoretical information

The main assumption of α_i model, described in detail in [9] introduced the **new measure – α_i coefficient** of reagent/lubricating oil or additive **properties / structure** influence on its **reactivity** related to **reaction conditions**. This model was worked out on the bases of the results of tribochemical investigations of different lubricating oils. In relation to tribological processes, **tribochemical reaction conditions** depend on the work done on tribological system - L. The work L is the function of applied load P, which can be treated as the only one variable y – L = f(y). The reagents **reactivity** is described by the second function of variable y - $\phi(y)$. Basing on Cauchy theorem the relation between two functions of the same one variable y is as follows:

$$\alpha_i = \{[f(b) - f(a)] / [\phi(b) - \phi(a)]\} / [f'(b) / \phi'(b)] \tag{3}$$

where *a* and *b* are values of y parameter. For different reagents value *b* is the only one variable (assumption of Cauchy theorem) and *a* is constant. Variable *b* was related to tribological process conditions – for example applied load P. In α_i model P is critical value and it relates either to seizure load or weld load. Consequently work done on the system means the work needed to achieve the seizure or weld.

Work done on tribological system can be related to thermodynamic description of tribological process. In the consequence reactivity can be related to the internal energy change Δu, and Eqn. (4) can be written as follows:

$$\alpha_i = [(L - L_o) / (\Delta u - \Delta u_o)] \times (d\Delta u / dL) \tag{4}$$

$L = \mu P v t$; μ - friction coefficient; v – speed; t – time; P – applied load (test result);
where: $L_o = f(a)$, $\Delta u_0 = \phi(a)$

Since it is difficult to define relationship $\Delta u = \phi(P)$ because Δu is not linear dependence of the applied load P, we can use the first low of thermodynamics L to express it as a function of Δu:

$$L = Q + \Delta u \tag{5}$$

Q is energy dissipated by system during tribological process; mainly it is a dissipated heat, which in relation to tribological process can be described by the following dependence:

$$Q = c_h (T - T_{ot}) \tag{6}$$

$$T = T_b + A P^{0,5} \tag{7}$$

where: c_h - average specific heat capacity,
T_b – temperature of lubricant out of friction area, T_{ot} – temperature of environment.

Assuming that both average specific heat capacity and T_b are constant for different oils, Q can be expressed as

$$Q = A_1 P^{0,5} \tag{8}$$

Consequently

$$\alpha_i = (\mu Pvt - \mu P_o vt) [(\mu vt\, dP - A_1\, dP^{0,5}) / \mu vt\, dP] / (\mu vtP - A_1 P^{0,5} - \Delta u_o) \tag{9}$$

3.2. Description of the C value as the function of applied load P in terms of Figure 1

It was experimentally found during tribological tests, described in [9], that α_i is linear function of applied load P. This linear dependence takes place in case the number of lubricants containing additives of different activity, but similar chemical structures were tested in the given tribological test. In this case

$$\alpha_i = (\mu Pvt - \mu P_o vt)\, C \tag{10}$$

and for additives of similar chemical structure (for example zinc dithiophosphates) C value is constant. For these additives in the given tribological test

$$C = [(\mu vt\, dP - A_1\, dP^{0,5}) / \mu vt\, dP] / (\mu vtP - A_1 P^{0,5} - \Delta u_o) \tag{11}$$

The fact that for additives of similar chemical structure (for example zinc dithiophosphates) C value is constant and C is the function of P, which value is different for each additive, requires that C is a harmonic function of P, described by the following dependence:

$$C = A \exp(-ay) / \cos(by + d) \tag{12}$$

where y = P or T and A, *a*, *b* and *d* are constant value.

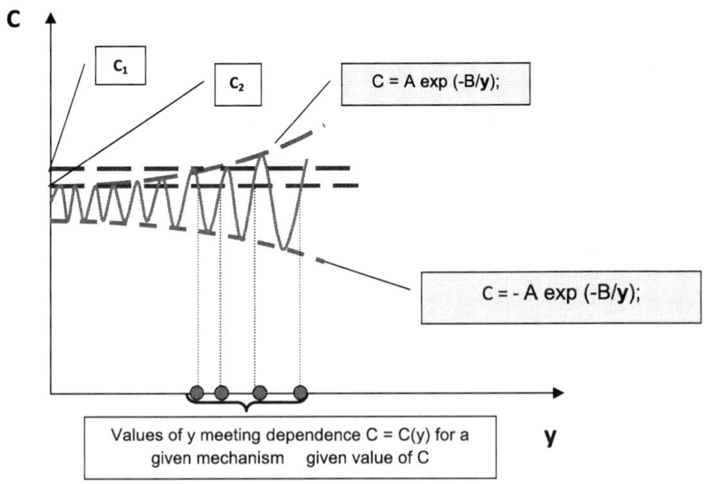

Figure 1. Dependence between C and applied load P

Consequently for the given constant value of C only some of values P meet equation (10) and (11). (red points in Figure1.)

What is physical and chemical meaning of those dependences? Answers seem to be also related to tribocatalysis and heterogeneous catalysis. Exponential part of this dependence can be connected with kinetics of chemical reactions by Arrhenius equation:

$$k = A \exp (B - E_a/RT) \tag{13}$$

where temperature T may be connected with load P applied in tribological process, as earlier shown in equation (7)

$$T = T_b + \mu P^{0,5} D$$

$D = 10^{-5} V_s P_p^{0,5} (k_1 + k_2)^{-1}$ where: V_s – sliding speed, P_p - unit pressure of the metal flow, k_1 and k_2 - coefficients of thermal conductivity of cooperated elements of tribological system [10].

3.3. Cauchy`s theorem application to the model

The α_i model and the equation (12) combine the activation energy delivered to molecules from the energy stream. Considering the energy delivered to molecules for each reaction, the following should be noted:

- Cauchy`s theorem requires the only one variable in functions f(y) and ϕ(y); application of α_i model to tribochemistry/ tribocatalysis and heterogeneous catalysis this variables can be temperature T or connected with them applied load P (tribochemistry/ tribocatalysis).
- To use α_i model in tribochemical and chemical problems solving it was assumed that variable y describes critical stage of reaction system, which is equivalent to critical rate of described reactions.
- A critical rate of catalytic reactions or triboreactions is reached at a critical temperature (eg. temperature of catalytic reaction at which there is maximum rate of this reaction) or, equivalently, under a critical load, at which seizure load takes place.
- Connecting critical rate of triboreaction / catalytic reaction with C in Eqn. 9 and 10 it was concluded that the value of E_a is the same for different reactants in the case of the same mechanism of triboreactions / catalytic reactions, because E_a is neither a function of *T* nor *P* and Cauchy`s theorem does not accept another variables than y.

Concluding for tribocatalysis:

- C is the function of P only, so activation energy E_a has to be constant for different additives / reactants and different values of critical load P.
- Looking on mathematical analysis for given mechanism there are permitted values of critical load, and consequently permitted values of critical rate of tribochemical reactions.

On the basis of this mathematical model, physical model of tribocatalysis / catalysis was created. The model assumes that mechanical work done on the system (containing liquid reagents = lubricants), is transformed to internal energy increase and dissipated energy. Internal energy is distributed in the system: one part is distributed to the liquid phase and is responsible for ambient temperature increase of the lubricant (T_a), the second part is cumulated in solid body (solid elements of tribological system) and is emitted as electrons or photons by its surface as impulses of high intensity. Energy cumulated in the liquid phase is not sufficient to reach value of E_a. Energy emitted by surface as impulses can reach value of E_a and reaction / triboreaction begins to proceed.

Eqn. 10 points that value of C depends not only on reaction rate constant but on intensity of energy emitted in angle γ from solid body to reaction space. Angle γ can not be another variable in Eqn. 8 and 10 – requirements of Cauchy`s theorem, so it should depend on T or P. Accordingly, the value of the angle γ ◎depends on the system energy flux. The critical state of a tribological system appears at conditions resulting in destruction of the protective film. It has been observed that for different reactants, the critical rate of reaction leading to protective layer destruction was achieved for different values of energy flux into the system (different values of applied load *P*). The same value of C obtained for different reactants and different densities of energy streams introduced into the system (characteristic of each reactant) leads to the conclusion that the same critical rate of destruction reaction was achieved and thus for each reactant a different angle γ is connected with the critical rate of reaction (different values of T or P).

Consequently, for each reactant there is a specific value of the energy flux density (e_γ), where $e_\gamma = e_0 \cos \gamma$, e_0 - intensity of energy stream emitted by the solid/catalyst in normal direction to solid surface ($\gamma = 0$) and the value of activation energy E_a is constant. Accordingly, it is possible to emphasize that:

i. E_a is constant for a given type of reaction and the critical rate of reaction depends not only on the energy quantity added to reactants but on the density of the introduced energy stream (time of the tribological process is constant for each load);
ii. The catalyst emits pulses of energy flux of high density at an angle γ. The value of emitted in short time energy is equal to the difference between activation energy calculated for reaction without catalyst E_a and activation energy calculated for catalytic reaction activation energy E_{ac};
iii. The catalyst collects energy done as mechanical work and emits it as pulses of the high energy density flux, thereby decreasing temperature needed for the reaction initiation or its rate enhancement.

3.4. Empirical verification of α_i model and its derivative equation concerning heterogeneous catalysis

3.4.1. *Verification of α_i model in tribological tests*

Basing on the above dependences and obtained test results the α_i values were calculated for a number of gear, hydraulic and transmission oils of viscosity at 40 ^{0}C in range between 32 to 220 mm^2/s. Because of the structure of protective layer created by lubricant under boundary conditions is selected on AW and EP types. The α_{AW} and α_{EP} coefficients are assigned to each tested lubricants. The method used in calculation of α_{AW} and α_{EP} values on the bases of tribotests results for each lubricant was described in [9].

The α_i values assigned to those lubricants were related to the results of tribological standard tests. Therefore, it was necessary to find out tribological experimental methods able to provide test results concerning lubricant's ability to create both AW and EP types of the boundary layer. In the already cited work [5], there were selected two different tests carried out using a four ball machine.

The first test was used to determine seizure load P_t under the following operating conditions:

- rotating speed of upper ball 470 +/- 20 rpm
- load increase continuously from 0 to the seizure load
- seizure load is detected by the friction coefficient measurement.

Following dependence for α_{AW} determined applying the above described procedure and P_t values is found:

$$\alpha_{AW} = (0{,}000086\ \nu_{40} - 0{,}01)0{,}5\ \mu\ P_t^2\, v_s\, v_P^{-1} + 0{,}2 - 0{,}00073\ \nu_{40} \tag{14}$$

where:

ν_{40} – kinematic viscosity of tested oil at 40 °C
v_s - sliding velocity (0,18 m/s)
v_P - speed of load increase (45 N/s)
μ- friction coefficient

This dependence points out that the mechanism of AW type of protective layer formation depends on base oil viscosity.

The α_{EP} values determined using above procedure were related to the second 4-ball test, where welding load P_w was the test result. This test was performed under the following operating conditions:

- rotating speed of upper ball 1470 +/- 30 rpm
- stepwise load increase according to Polish Standard PN-76/C-04147 (ASTM D2783: Standard Test Method for Measurement of EP Properties of Lubricating Fluids, Four-Ball Method)
- test duration on each step 10 s

It was found that P_w values depend on durability of EP type of the boundary layer and was experimentally found the linear dependence between α_{EP} and P_w:

$$\alpha_{EP} = 0{,}48 - 0{,}00013\ P_w \tag{15}$$

Using both these standard tests the value of α_{AW} and α_{EP} can be determined easily and than used to predict the results of another tribological tests (eg. FZG) or analyze the structure of protective layer created in various machines (participation of AW and EP structure in protective layer).

3.4.2. Verification of derivative equation concerning heterogeneous catalysis; electron and photon emission anisotropy

Eqn. 12 (C = A exp(-*a*y) / cos (*b*y + *d*)) leads to conclusion that the emission of energy (electrons and photons) from the surface of solid body is anisotropic one. Exhaustive literature review revealed that there is no data on the angular distribution of triboemitted electrons, but that existing research works on the emission of electrons from cathodes consistently report anisotropic distributions [7, 11-13]. Highly anisotropic distributions with a maximum in the direction normal to the emitting surface were measured.

Figure. 2. illustrates an example of anisotropic electron emission from sandwich cathodes. At the emission temperature of 300K the measured electrons present a quasi-isotropic characteristic. But at a lower temperature of 80K, the isotropic component vanishes.

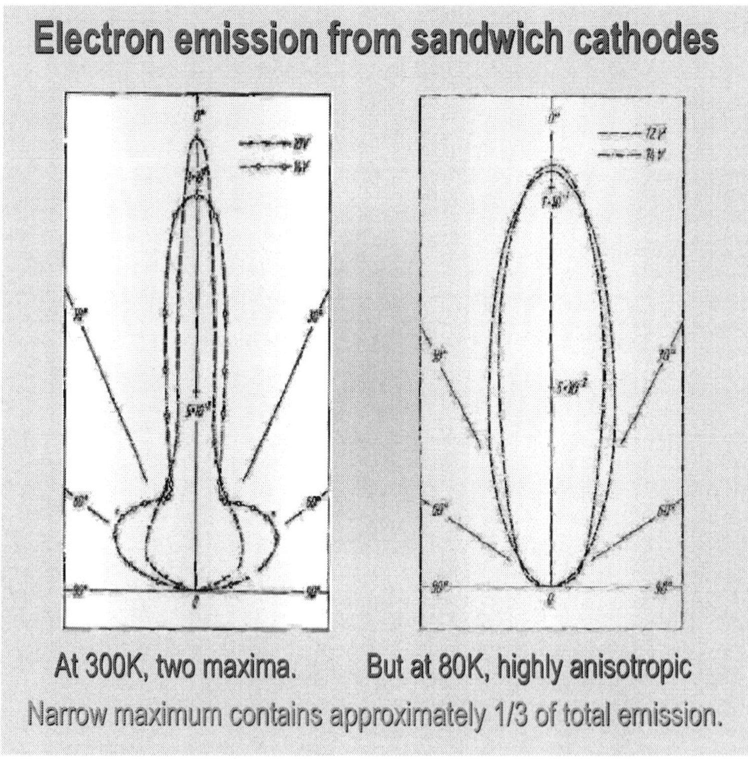

Figure 2. Examples of EE from sandwich cathodes presenting 2 maxima with a maximum in direction normal to the emitting surface, and highly anisotropic EE: based on Hrach research results [7,13]

References [11-12] allow to make an estimate that at room temperature the narrow maximum around the direction normal to the emitting surface contains approximately 1/3 of the total emission. Further work of Hrach [7,13] was of particular significance as he measured the energy characteristics of emitted electrons at different angles by means of a hemispherical collector and the retarding-grid technique. He found that at room temperature (300K) anisotropic energy spectra of emitted electrons were in the range of 0 and 7eV, but for emission angles closer to the normal to the surface the measured energy was between zero and 4 eV. Figure 3. depicts some of these findings.

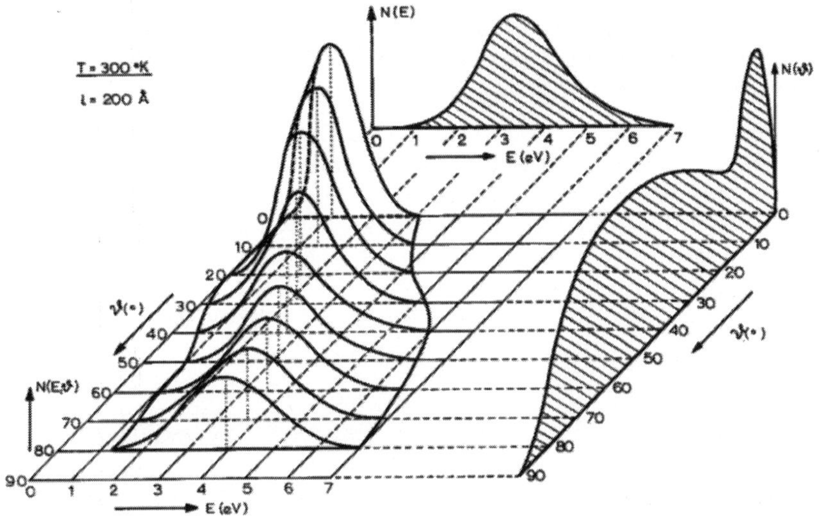

Figure 3. Typical experimental angular and energy distribution of electrons emitted from Al-Al_2O_3-Au sandwich cathode structures. Temperature: 300K, applied voltage: 10V [13]

Reference [3] includes more detailed information concerning figures 2 and 3. More recent study [14] was on the angular distribution of thermo-stimulated exoelectron emission (TSEE) from alpha-Al_2O_3. Anisotropic angular distributions were strongly directed normally to the surface; the energy of the emitted electrons was in the range of 0 to 4 eV. Jablonski and Zemek [8] found highly anisotropic distribution similar to one of electron-type for the photon emission from X-ray-irradiated thin polycrystalline aluminum foils.

3.5. Disscussion on the basic Arrehnius equation

In Arrhenius equation T equals T_a, ($T = T_a$) and it relates to ambient temperature of reaction mixture. Conclusions from α_i model are as follows:

- When catalyst is used ambient energy RT_a is less than energy in space near catalysts particles surface = RT_s. T_s is the calculated temperature near catalyst surface and it should be higher than ambient temperature of reaction mixture. RT_s is real energy acted to the molecules of reactants and this energy should be introduced into Arrhenius equation.

$$k = \acute{A} \exp (B - E_a / RT_s) \tag{16}$$

Relation between RT_a and RT_s is as follows:

$$\exp (B - E_a / RT_s) / \exp (B - E_a / RT_a) = 1 / (e_o \cos \gamma) \tag{17}$$

$$E_a / RT_s = E_a / RT_a + \ln (e_o \cos \gamma) \tag{18}$$

When ($e_o \cos \gamma$) > 1 the real energy near catalyst surface is less than ambient energy ($RT_s < RT_a$) – reaction inhibition, when ($e_o \cos \gamma$) < 1 the real energy near catalyst surface is higher than ambient energy ($RT_s > RT_a$) – catalytic effect, for ($e_o \cos \gamma$) = 1 there is no catalytic nor inhibitor effect.

In case the same value of reaction rate constant for reaction without and with catalyst is compared using Arrhenius equation, there should be noticed difference between activation energy for reaction with catalyst (E_{ac}) and without catalyst (E_a). The hypothesis based on α_i model is that this difference is equal to energy emitted by solids surface in angle γ.

$$\Delta E_a = 40 \text{ kJ/mol} = 240 \times 10^{22} \text{ eV/mol} = 4 \text{ eV / molecule}$$

Empirically determined energy emitted by solid surface is in the range 3 to 7eV and it is the range of value of ΔE_a.

The results of these calculations are in line with hypothesis based on α_i model saying that catalytic effect is due to energy emission from catalysts surface in the form of electrons / photons stream, additional energy of which makes possible to reach the same reaction rate in lower ambient temperature or increase the reaction rate in the same ambient temperature. This hypothesis is described by C in Eqn. 8: C is the quotient of reaction rate constant described by Arrhenius equation and the stream of energy emitted by the surface of catalyst in angle γ.

Concluding the general dependence (1):

$$\alpha_i = \{[f(b) - f(a)] / [\phi(b) - \phi(a)]\} / [f'(b) / \phi'(b)] \tag{1}$$

can be shown as follows:

$$\alpha_i = (L - L_0) A [\exp (B - E_a/RT_a)] / e_0\cos(\gamma) \tag{19}$$

therefore, it is concluded that:

$$\{[f(b) - f(a)] / [\phi(b) - \phi(a)]\} / [f'(b) / \phi'(b)] = \\ = (L - L_0) A [\exp (B - E_a/RT_a)] / e_0\cos(\gamma) \tag{20}$$

Function f(y) represents the stream of energy introduced into the system, function $\phi(y)$ is connected with catalytic / tribocatalytic reaction critical rate, which is represent by the ratio of reaction rate constant described by Arrhenius equation and the stream of energy emitted by catalysts surface in angle γ. This ratio points that to the equation describing reaction rate constant (Arrhenius equation) should be added denominator describing the stream of energy emitted by catalysts surface. The reaction rate constant described by the above ratio leads to another explanation of the mechanism of catalytic effect than, based on Arrhenius equation, decreasing of the value of activation energy.

3.6. Practical significance of the new (E_a) approach in tribo- and mechano-chemistry

Mechanochemistry is the coupling of the mechanical and the chemical phenomena on a molecular scale and includes mechanical breakage, chemical behaviour of mechanically-stressed solids (e.g., stress-corrosion cracking), tribology, polymer degradation under shear, cavitation-related phenomena (e.g., sonochemistry and sonoluminescence), shockwave chemistry and physics, and even the burgeoning field of molecular machines. Mechanochemistry can be seen as an interface between chemistry and mechanical engineering. A smart method was proposed recently, in order to measure the energy involved during mechanical transformations. Displacement reactions between a metal oxide and a more reactive metal can be induced by ball milling. In some cases the reaction progresses gradually and a metal/metal-oxide nanocomposite is formed. Ball milling may also initiate a self propagating combustive reaction. The information available about these processes is reviewed. It is argued that the gradual or combustive nature of the reaction depends on thermodynamic parameters, the microstructure of the reaction mixture, and the way they develop during the milling process.

Baláž, et al. [15] investigated the mechanochemical treatment of solids which lead to a positive influence on the solid – liquid kinetics. They used Arrhenius equation for activation energy analysis. The breaking of bonds in the crystalline lattice of solids brings about a decrease (ΔE^*) in the activation energy and an increase in the rate of leaching

$$(\Delta E^*) = E - E^* \tag{21}$$

$$k^* = k \exp (\Delta E^*/RT) \tag{22}$$

where E is the apparent activation energy of the non-disordered solid, E^* is the apparent activation energy of the disordered solid, k, R and T stand for the rate constant of leaching for the non-disordered solid, (the pre-exponential factor) gas constant and reaction temperature, respectively; k^* is the rate constant of leaching for the disordered solid. If $E > E^*$, then $\exp (\Delta E^*/RT) > 1$ and thus it follows from Eqn. (18) that $k^* > k$, i.e., the rate of leaching of a disordered solid is greater than that of an ordered mineral.

Thermodynamic methods are essentially macroscopic by origin and nature. They appear in the analysis of macroscopic engineering systems. They have been reliably validated in numerous macroscopic experiments and observations. Most probably there can be found areas that permit analysis of mechanochemical systems by means of relatively simple thermodynamic methods. From the purely thermodynamic point of view, the central problem of mechanochemistry is the exchange of energy between the (long-range) elastic energy and the (short-range) energy accumulated in individual bonds.

There is no clear theory which could be adapted to mechanochemistry, however the most recent approach [16] should be mentioned here. The α_i model applied for tribochemical applications can also be adapted to mechanochemistry. This model can be helpful in general dependences formulation, related to kinetics of mechanochemical reactions and to mechanical forces used for reactions activation. The theory based on α_i model assumes that mechanical energy introduced into solid body – reagent or catalyst, is accumulated in it and then emitted as low energy electrons or photons of energy equal or higher than activation energy of the reaction. The general Eqn. (19) can be used to determine quantitatively relationship between mechanical stress (L), the possibility of solid body to accumulate and then emit energy ($e_0 \cos \gamma$) and kinetics of mechanochemical reaction.

Mechanochemistry, especially results of investigations shown above can be explained by hypothesis based on α_i model. On the other hand the positive effect of mechanical stress on catalyst efficiency confirms this hypothesis. However the reason of this effect can be mechanically (eg. during milling) produced changes of catalysts surface.

Rodriguez et al. [17] found out the influence of ultrasound radiation on catalysts effectiveness. They tested a new advanced method for dechlorination of 1,2,3-, 1,2,4-, and 1,3,5-trichlorobenzenes in organic solvent catalysed by palladium on carbon support and solid hydrazine hydrochloride yields benzene in short reaction times. The catalyst system can be efficiently reused for several cycles. Ultrasound radiation of the heterogeneous catalyst reaction increases remarkably the rate of dechlorination. Moreover, Rodriguez found that there is optimum energy of ultrasound radiation which results maximum catalysts efficiency. This effect is not seen when ultrasound radiation act liquid reactants. Rodriguez results confirm thesis that energy, in this case of ultrasound radiation is useful in reaction rate increasing when solid body – particles of catalyst are present in reaction mixture. This energy is cumulated by catalyst and emitted to the space near catalyst surface, what is the reason of reaction rate increasing. This effect can not be explained by the changes of the structure of catalysts surface, like in other mechanochemical effects (eg. during milling) so the only one probably mechanism is emission of cumulated in solid catalyst energy to reaction space.

The concept of mechanochemistry to modify molecular reactivity has a rich history for a long time. For instance Kauzmann an Eyring as early as 1940 [18] suggested that the mechanical perturbation of diatomic molecules could alter the reaction coordinates combined with their homolytic dissociation.

The chemical kinetic quantitatively describes homogenous reactions, where the rate of reaction depends only on heat introduced to reaction system. Kinetic equations concern reagents concentration, and according to Arrhenius equation: temperature of reaction mixture as well as activation energy.

These kinetic equations used in heterogeneous catalytic reactions description concern no one parameter characterizing catalyst. Consequently the effect of catalyst action can be explained only by the decreasing of activation energy value – the only one calculated parameter in kinetic equations. It is the reason that all theories of catalysts action try explain the mechanism of activation energy decreasing.

The α_i model, particularly equation (19) describes catalytic and tribocatalytic reactions by dependence concerning parts, which quantitatively characterizes:

- all kinds of energy introduced into the reaction system, including mechanical energy - L in eqn. (19)
- properties of catalyst, explained by energy emitted from its surface to reaction space – $e_0 \cos \gamma$.

Resulted from the α_i model concept of the mechanism of heterogeneous catalysis and tribocatalysis, shown above, is confirmed partly by tribochemistry and mechanochemistry. This mechanism should be directly confirmed, particularly by materials engineering, which should explain:

- the role of support in catalysts activity,
- the possibility of energy storage by different materials
- the influence of catalyst surface structure on the electrons / photons emission

Most recent work [19] emphasizes that while the detailed mechanisms by which different mechanochemical phenomena arise are not always well understood, mechanical forces are capable of effecting novel reactivity. Additionally, it strengthens that using force, one can effectively shepherd a chemical reaction down specific reaction pathways, for instance *by selectively lowering the energy of a transition state.* At this point it is of note, that the field of polymer mechanochemistry, has also the potential to change this paradigm by revolutionizing the way chemists think about controlling chemical reactions [19].

In recent years the mechanochemistry field approach has found a renaissance, and different techniques have been applied to activate chemical reaction [20-22] and thereby to lower their activation energy.

4. Conclusions

i. The α_i model put forward in this paper attempts to correlate mechanical work performed on a solid with its catalytic activity. This model was worked out on the basis of tribological tests results and was dedicated to tribochemistry.
ii. The analysis of basic dependences resulted from α_i model lead to the conclusion that the mechanism of catalysis related to tribological processes can be adapted to heterogeneous catalysis including mechanochemical reactions.
iii. The reactants molecules energy resulted from ambient temperature of reaction mixture is enhanced near solid surface by additional energy – emitted electrons / photons. Due to the additional energy the reaction can reach a critical rate.

iv. The energy emitted from surface as pulses ranges 3–5 eV and can reach the value of activation energy (E_a) and the triboreaction process starts to proceed or reaches the critical rate.

v. Based on the discussion concerning the α_i model, thermionic emission, and the NIRAM approach it is concluded that for both thermochemical heterogeneous reactions and catalyzed heterogeneous processes, the same activation energy (E_a value) is needed to initiate the reaction process.

vi. The hypothesis, based on α_i model is that the mechanical work done on the reaction system is transformed to internal energy increase.

vii. The internal energy is distributed into a liquid / fluid phase bringing about ambient temperature increase (T_a), and a part of introduced energy is accumulated in solid machine elements of tribological systems and, catalyst particles in catalytic processes.

viii. This accumulated energy is emitted by solid surface to reaction space as electrons / photons. The electron / photon emission is anisotropic one. There is a specific angle γ at which emitted energy is suitable to activate reactant molecules.

In the summary it can be said that the problem of Arrhenius equation adaptation to heterogeneous catalysis as well as tribocatalysis might be solved using α_i model. Instead of Arrhenius equation in reaction rate description should be used the quotidian of reaction rate constant according to Arrhenius equation and the stream of energy emitted by the surface of catalyst in angle γ. The reaction rate constant described by the above ratio leads to another explanation of the mechanism of catalytic effect than, based on Arrhenius equation, decreasing of the value of activation energy. This effect is due to addition portion of energy emitted by catalysts surface to the reaction space. At this point it should be emphasized that equation (19) describes both catalytic and tribocatalytic reactions. This equation quantitatively characterizes all kinds of energy introduced into the reaction system, including mechanical energy and properties of catalyst, explained by energy emitted from its surface to reaction space.

Author details

Andrzej Kulczycki
Air Force Institute of Technology, Warsaw, Poland
Cardinal Stefan Wyszynski University, Warsaw, Poland

Czesław Kajdas
Warsaw University of Technology, Institute of Chemistry in Plock, Poland
Automotive Industry Institute PIMOT, Warsaw, Poland

5. References

[1] Bond G.C. Heterogeneous Catalysis: Principles and Applications. Oxford: Clarendon Press, 1987.

[2] Kajdas C.K, Kulczycki A. A New Idea of the Influence of Solid Materials on Kinetics ofChemical Reactions. Materials Science – Poland 2008, 26, 787 - 796. http://materialsscience.pwr.wroc.pl/bi/vol26no3/articles/ms_2008_375.pdf

[3] Kajdas C.K, Kulczycki A, Kurzydłowski K.J, Molina G.J. Activation Energy of Tribochemical and Heterogeneous Catalytic Reactions. Materials Science-Poland 2010, 28, 523-533. http://materialsscience.pwr.wroc.pl/bi/vol28no2/articles/ms_15_2009-373kajdas.pdf

[4] Dante R.C, Kajdas C.K, Kulczycki A. Theoretical Advances in the Kinetics of Tribochemical Reactions. Reaction, Kinetics Mechanisms and Catalysis 2010, 99, 37 – 46.

[5] Klier K. The Transition State in Heterogeneous Catalysis, Topics in Catalysis 2002, 18, 3-4, 141 -156.

[6] Sato N, Seo M. Chemically Stimulated Exo-emission from a Silver Catalyst. Nature 1967; 216(Oct) 361-362. DOI:10.1038/216361a0

[7] Hrach H. Energy Angular Distribution of Electrons Emitted from MIM Systems. Thin Solid Films 1973, 15, 65-69. http://dx.doi.org/10.1016/0040-6090(73)90204-6

[8] Jablonski A, Zemek J. Angular Distribution of Photoemission from Amorphous and Polycrystalline Solids. Physical Review B, 1993, 48, 4799-4805. DOI:10.1103/PhysRevB.48.4799

[9] Kulczycki A. The Correlation Between Results of Different Model Friction Tests in Terms of an Energy Analysis of Friction and Lubrication. Wear 1985, 103, 67-75.

[10] Hebda M, Wachal A. Trybologia, Warszawa: Wydawnictwa Naukowo-Techniczne, 1980.

[11] Hrach R, May J. Electrons Emitted from MIM Structures. Physica Status Solidi (a), 1977, 1, 637-642.

[12] Gould R.D, Hogarth C.A. Angular Distribution Measurements of Electrons Emitted from Thin Film Au–SiO_x–Au Diode and Triode Structures. Physica Status Solidi (a), 1977, 41, 439-442. DOI: 10.1002/pssa.2210410212

[13] Hrach R. Emission of Electrons from MIM Systems: Disscussion of Processes in the Cathode. Czechoslovak Journal of Physics B, 1973, 23(2), 234-242. http://link.springer.com/article/10.1007/BF01587248?null

[14] Fitting H.-J, Glaefeke H, Wild W, Lang J. Energy and Angular Distribution of Exoelectrons. Physica Status Solidi (a), 1977, 42, K75–K77.

[15] Baláž P, Choi W.S, Fabian, M, Godocikova E. Mechanochemistry in the Preparation of Advanced Materials. Acta Montanistica Slovaca 2006, 11(2), 122-129.http://actamont.tuke.sk/pdf/2006/n2/5balaz.pdf

[16] Hiratsuka K, Kajdas C. Mechanochemistry as a Key to Understand the Mechanisms of Boundary Lubrication, mechanolysis and Gas Evolution during friction. Proc. IMechE Part J: J. Engineering Tribology, 2012 (submitted for publication).

[17] Rodriguez J.G, Lafuente A. A New Advanced Method for Heterogeneous Catalysed Dechlorination of 1,2,3-, 1,2,4-, and 1,3,5 – Trichlorobenzenes in Hydrocarbon Solvent. Tetrahedron Letters 2002, 43, 9645 – 9647. http://144.206.159.178/ft/1010/73973/1269348.pdf

[18] Kauzmann W, Eyring H. The Viscous Flow of Large Molecules. Journal of American Chemical Society 1940, 62, 3113-3125.

[19] Wiggins K.M, Brantley N, Bielawski C.W. Polymer Mechanochemistry: Force Enabled Transformations. American Chemical Society Macro Letters 2012, 1, 623-626. DOI: 10.1021/mz300167y

[20] Beyer M.K, Clausen-Schaumann H. Mechanochemistry: the Mechanical Activation of Covalent Bonds. Chemical Reviews 2005; 105(8) 2921-2948. DOI:10.1021/cr030697h

[21] Caruso M.M, Davis D.A, Shen Q, Odom S.A, Sottos N.R, White S.R, Moore J.S. Mechanically-Induced Chemical Changes in Polymeric Materials. Chemical Reviews 2009, 109, 5755-5798. DOI: 10.1021/cr9001353

[22] Kaupp G. Mechanochemistry: the Varied Applications of Mechanical Bond-Breaking, The Royal Society of Chemistry, CrystEngComm 2009, 11, 388–403. DOI: 10.1039/b810822f

Chapter 5

Solid Particle Erosion on Different Metallic Materials

Juan R. Laguna-Camacho, M. Vite-Torres, E.A. Gallardo-Hernández and E.E. Vera-Cárdenas

Additional information is available at the end of the chapter

http://dx.doi.org/10.5772/51176

1. Introduction

Testing on ferrous and non-ferrous materials has been widely carried out to study their erosion resistance. Venkataraman & Sundararajan [1] conducted a study about the solid particle erosion of copper at a range of low impact velocities. In this particular case, the eroded surface was completely covered with the erosion debris in the form of flakes or platelets. These flakes appeared to be completely separated or fractured from the material surface and were flattened by subsequent impacts. For this reason, it was concluded that at low impact velocities the erosion damage was characterized mainly by lip or platelet fracture whereas it was distinguished with lip formation (rather than its subsequent fracture) at higher impact velocities.

Additionally, studies on the erosion behaviour of AISI 4140 steel under various heat treatment conditions was investigated by Ambrosini & Bahadur [2]. In this work, the investigation was concentrated on the effect of various microstructures and mechanical properties on the erosion resistance. A constant velocity of 50 m/s was used for all the erosion tests. The target was impacted at an angle of 30º to the specimen surface, the particle feed rate was 20 g/min, SiC particles, 125 μm in size, were used as the abrasive. From the results, it was concluded that erosion rate increases with increasing hardness and ultimate strength, but decreases with increasing ductility. In this particular work, the heat treatment with the optimum combination of erosion resistance and mechanical properties was oil quenching followed by tempering in the temperature range 480-595 ºC for 2 h. In addition, SEM studies presented severe plastic deformation in the eroded zones together with abrasion marks, indicating that material subjected to erosion initially undergoes plastic deformation and is later removed by abrasion.

Harsha & Bhaskar [3] carried out research to study the erosion behaviour of ferrous and non-ferrous materials and also to examine the erosion model developed for normal and oblique impact angles by Hutchings [4]. The materials tested were aluminium, brass, copper, mild steel, stainless steel and cast iron. They determined from the SEM studies that the worn surfaces had revealed various wear mechanisms such as microploughing, lip formation, platelet, small craters of indentation and microcracking.

In addition to these studies, Morrison & Scattergood [5] carried out erosion tests on 304 stainless steel. In this work, it was concluded from the SEM observations that similar morphologies for low and high impact angles could be observed in ductile metals when they were subjected to the impact of sharp particles. The surfaces displayed a peak-and-valley topology together with attached platelet mechanisms. In addition, the physical basis for a single-mechanism to erosion in ductile metals was considered to be related to shear deformations that control material displacement within a process zone for a general set of impact events producing at all impact angles. These events included indentation, ploughing and cutting or micromachining. In respect to the effect of the erodent particle shape on solid particle erosion, Hutchings showed differences in eroded surfaces due to a shape particle effect [6]. It was observed that the shape of abrasive particles influences the pattern of plastic deformation around each indentation and the proportion of material displaced from each indentation, which forms a rim or lip. More rounded particles led to less localized deformation, and more impacts were required to remove each fragment of debris.

Liebhard & Levy [7] conducted a study related to the effect of erodent particle characteristics on the erosion of 1018 steel. Spherical glass beads of four different diameter ranges between 53-600 μm and angular SiC of nine different diameter ranges between 44-991 μm were the erodents. The particle velocities were 20 and 60 m/s, an impact angle of 30º was used to conduct all the tests and the feed rate was varied from 0.6 to 6 g/min. The results showed that there was a big difference in the erosivity of the spherical and angular particles as a function of particle size. Angular particles generally were an order of magnitude more erosive than spherical particles. In addition, the erosivity of spherical particles increased with particle size to a peak and then decrease at even larger particle sizes. In respect to angular particle erosivity, it was increased with particle size to a level that became nearly constant with size at lower velocities, but increased continuously at higher particle velocities. Lower flow rates caused more mass loss than higher flow rates for both spherical and angular particles.

In this work, the performance of different metallic materials has been analyzed. The aim of this experimentation was essentially to know the behavior of these materials against solid particle erosion and compare their erosion resistance. In addition, the functionality of both, the rig and the velocity measurement method was evaluated.

2. Experimental details

2.1. Specimens

The materials employed to conduct the tests were 4140 and 1018 steels, stainless steel 304, 316 and 420, aluminium 6061, brass and copper. The test surface of each specimen was ground using SiC emery paper grade 1200. The average roughness (R_a) in each specimen before testing was 1 μm. The samples had a rectangular shape with dimensions of 50 x 25 mm^2 and 3 mm in thickness. The abrasive particle used was silicon carbide (SiC) of an angular shape, as seen in Figure 1, with a particle size of 420-450 μm [8]. Table 1 presents the chemical composition of the materials used in the erosion tests whereas Table 2 shows the hardness of the materials. Microhardness values were obtained by calculating an average value, 10 different points were measured. The applied load was 100gf.

Figure 1. Size and morphology of the abrasive particles [8]

Material	C	Si	Mn	Mg	P máx.	S máx.	Cr	Ni	V	Cu	Mo	Pb	Zn	Ti
4140	0.38-0.43	0.15-0.35	0.75-1.00	-	0.035	0.040	0.80-1.10	-		-	0.15-0.25		-	-
1018	0.15-0.20	0.15-0.35	0.60-0.90	-	0.040	0.50	-	-	-	-	-		-	-
304	0.08	1.00	2.00	-	-	-	16.00 - 18.00	8.00-10.50		-	-		-	-
316	0.08	1.00	2.00	-	-	-	16.00 - 18.00	10.00 - 14.00		-	2.00-3.00		-	-
420	0.38	0.40	0.45	-	-	-	13.60	-	0.30	-	-		-	-
Aluminium 6061	0.40-0.82	-	0.15 Máx.	0.80-1.20	-	-	0.04-0.35			0.15-0.40	-		0.25 Máx.	0.15 Máx.
Brass	-	-	-	-	-	-	-	-	-	55.84	-	0.05 Máx.	Rem.	-
Copper	-	-	-	-	-	-	-	-	-	87.66	-	0.05 máx.	Rem.	-

Table 1. Chemical composition of materials

Material	Vickers Hardness (HV)
4140	280
1018	230
AISI 304	160
AISI 316	150
AISI 420	200-240
Aluminium 6061	130
Brass	228
Copper	161

Table 2. Hardness

2.2. Test procedure

The apparatus used to carry out the erosion tests is similar to that presented in the ASTM G76-95 test standard [8-10]. Figure 2 shows a schematic diagram of the rig that was developed. In this device, the particles of silicon carbide (SiC) were accelerated from a nozzle by using a compressed air stream that caused them to impact the surface of the material.

The materials were eroded in a time period of 10 min, although each sample was removed every 2 min to determine the amount of mass lost. The specimen holder could be rotated to be impinged at different incident angles (30º, 45º, 60º, 75º and 90º). These angles were selected to evaluate the materials at both low and high impact angles and to determine if the behavior of these materials was similar to the conventional materials that were used in other erosion studies [1-4]. A particle velocity of 24 ± 2 m/s and a constant abrasive flow rate of 0.7 ± 0.5 g/min were used to reduce the effects of interaction between the incident particles and the rebounding particles. The reduction of this effect was usually accomplished at lower impact angles such as 30° and 45°, where the abrasive particles commonly impacted the material, slid along the surface, and then fell away. However, a greater level of interaction between the particles was observed at 90°. The measurements of particle velocity were carried out by using an opto-electronic system.

In all of the tests, specimens were located 10 ± 1 mm from the end of the glass nozzle. The nozzle had the following dimensions: a 4.7 mm internal diameter, a 6.3 mm external diameter and 260 mm length. The room temperature was between 35 and 40º C. The specimens were weighed using an analytical balance with an accuracy of ± 0.0001 g, before the start of each test and were removed every 2 min, cleaned in an ultrasonic cleaning device using ethanol and weighed again to determine the mass lost. Subsequently, micrographs of eroded surfaces were obtained using a scanning electron microscope (SEM) to analyze and identify the possible wear mechanisms involved.

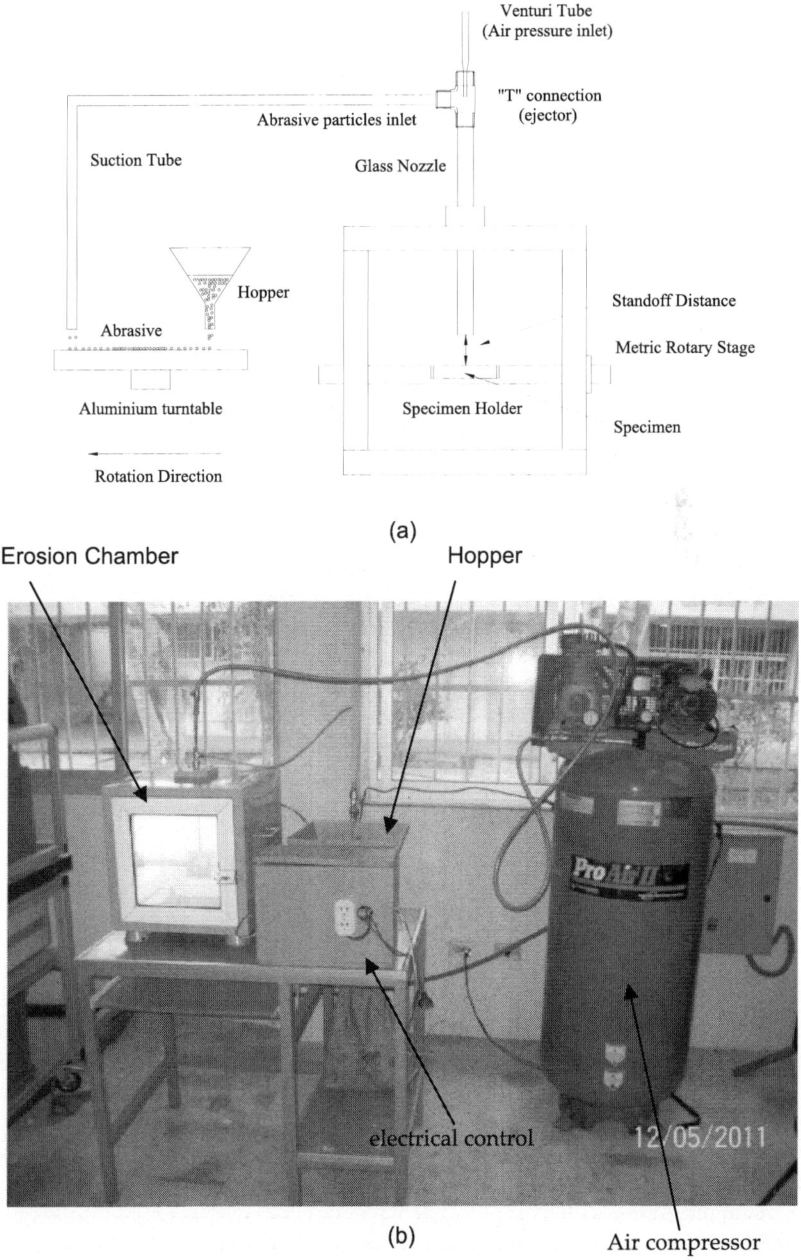

Figure 2. (a) Schematic Diagram of erosion rig developed, (b) Experimental setup [8]

2.3. Velocity measurement method

The particle velocity was measured with an opto-electronic flight-timer similar to that described by Kosel and Anand [11, 12]. This system offers the possibility to measure the particle velocity in an accurate mode and the design does not request high costs. It is practical and flexible in relation to the equipment that can be used to conduct the measurements, for instance, the oscilloscope, the emitters and detectors, etc. It must be mentioned because other equipment such as a high speed camera are expensive and result more difficult to obtain them to conduct a research project. The uncertainties of this particular system are that the measurements are not as consistent as expected. In certain cases, it is necessary to use more abrasive particles to obtain a signal and finally to complete the measurement process. In this specific work, several air pressures, from 0.35 kg/cm^2 (5 psi) to 3.86 kg/cm^2 (55 psi) were used to conduct the tests. It was concluded that a higher pressure and therefore a higher particle velocity gave better results. Due to this fact, 3.86 kg/cm^2 (55 psi) equivalent to 24 ± 2 m/s was the value chosen to carry out the erosion tests. Signals more consistent and clear were obtained using this particle velocity.

The velocity measurement method is mainly composed of two infra-red emitters and detectors held by a rectangular plastic block as observed in the schematic diagram in Figure 3a. A few abrasive particles pass through a glass tube attached at the end of the nozzle and are recognized by both light beams producing the signals that are processed through an amplification system connected to an oscilloscope. It is possible to determine the real time travelled by a few abrasive particles using the two signals, 1 and 2. The standard distance between signals was 10 ± 1 mm. Flight time data were continuously collected and stored. An average particle velocity was reached after 20 measurements. The set-up developed is shown in Figure 3b. As observed, the infra-red emitters were set on each side to ensure that most of the abrasive particles were monitored when passing through the glass tube. Finally, an example of the signals received directly in the oscilloscope is presented in Figure 3c.

3. Test results

3.1. Wear mechanisms

In Figure 4, it is possible to observe the wear scars obtained for all the tested materials at different incident angles. The wear scar area is reduced as the impact angle is increased. It has an elliptical shape at 30° and 45º whereas a roughly circular shape is observed at 60º and 90°. This can be related to the impact geometry which modifies the orientation of the specimen when it is positioned at different incident angles. In the schematic diagram shown in Figure 5a and b, A, represents the wear scar independently of the halo effect distinguished by the dashed lines. Due to this fact, the wear scar commonly shows an elliptical shape at incidence angles lower than 45º due to a higher divergence of the particle stream. A circular shape is often observed at higher impact angles near or at 90º as observed commonly in other erosion studies [13-17]. In both cases, the plume of abrasive particles is concentrated in the central part of the stream. All materials showed clearly the halo effect

[18], mentioned above, which is represented by a secondary erosion damage zone. The estimated area of the erosive scars including the halo effect is presented in photographs in Figure 4.

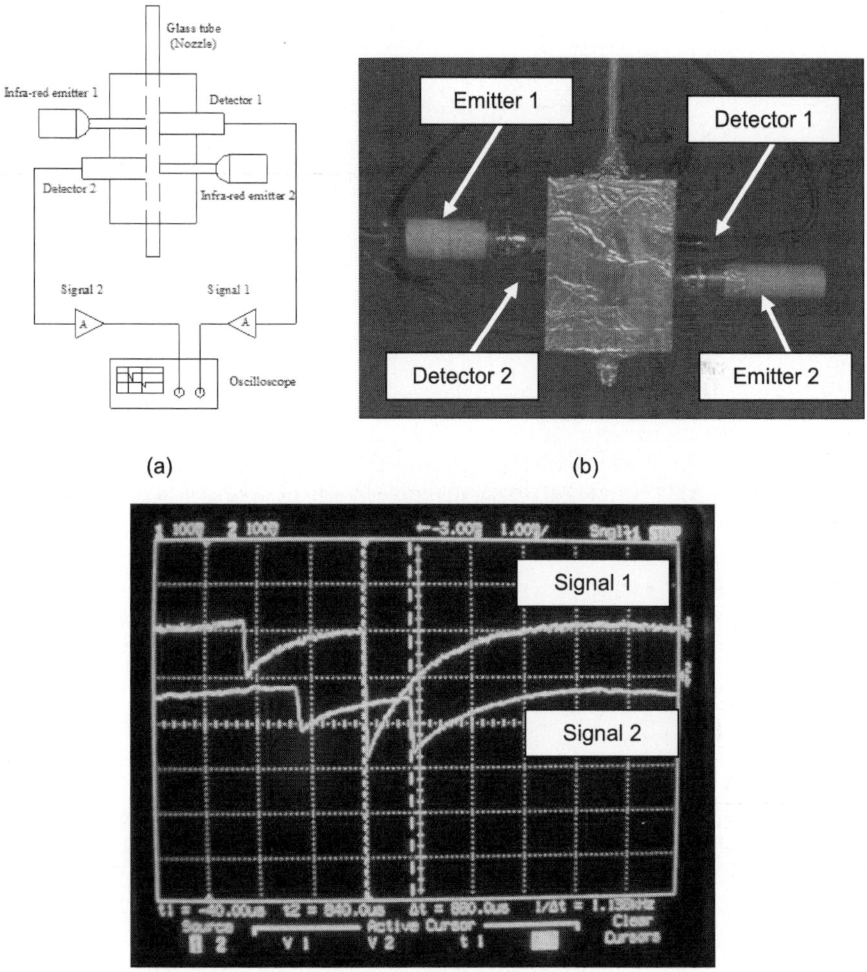

Figure 3. (a) Schematic diagram of set-up, (b) Real set-up developed, (c) Signals 1 and 2 processed in the oscilloscope

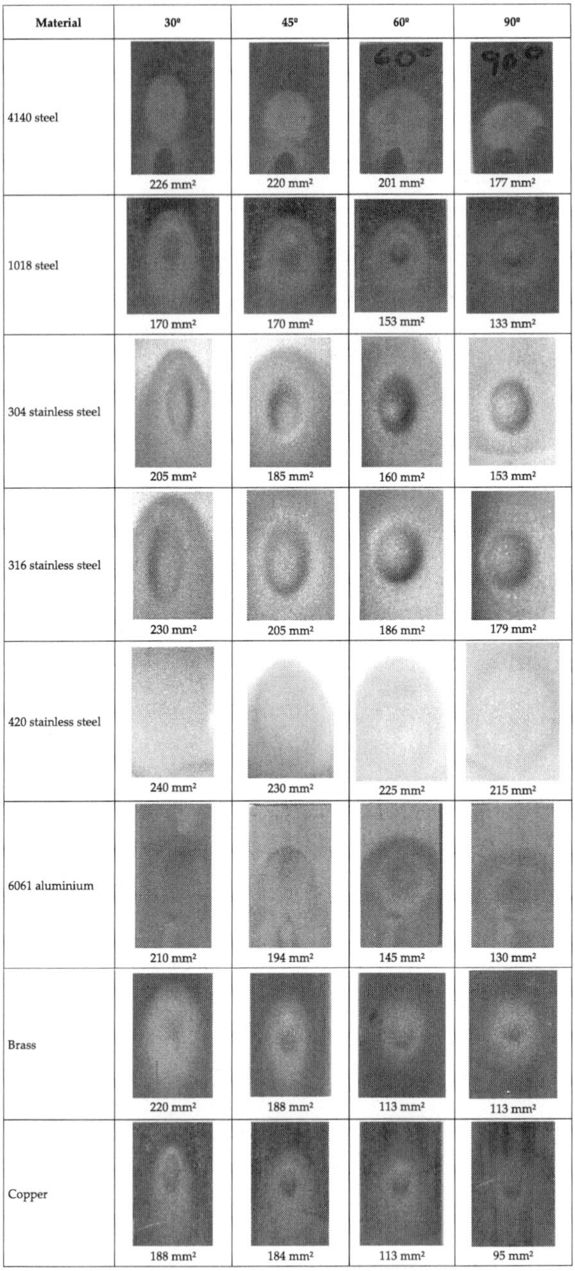

Figure 4. Erosion Damage on tested materials

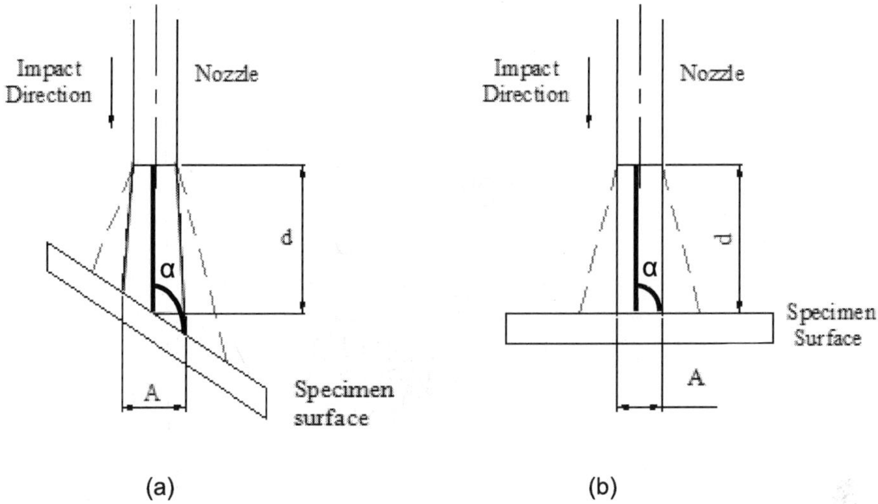

Figure 5. Impact geometry, (a) $\alpha \leq 45^{\circ}$, (b) Near or at 90º

Figure 6a-j present the wear mechanisms involved in this study on 4140 and 1018 steel and stainless steels 304, 316 and 420. In respect to AISI 4140, larger craters are clearly seen on the surface at 30° with some wear debris inside, whereas a more roughened surface is observed at 90°. AISI 1018 exhibited cracks located at random positions and also pitting action was seen on the surface at 45°. On the other hand, stainless steels 304, 316 and 420 showed pitting and ploughing action, irregular indentations, scratches and smeared wear debris at 30° and 45° whereas fragments of abrasive particles embedded or smeared on the surface were clearly seen at angles near or at 90°.

Photographs shown in Figure 7a- present the damage incurred on 6061 aluminium, copper and brass at different incident angles. Figure 7a shows the worn surface of aluminium; particles of aluminium debris and lips around the craters smeared by subsequent impacts of abrasive particles are observed as in other erosion studies [19-23]. Also, craters, pitting and striations are observed on the specimen surface. The reduction in mass loss at higher impact angles, near or at 90º, is because there was not too much evidence of sliding action of abrasive particles unlike lower impact angles where the sliding component is significant and increases the mass lost in the material. In addition, although a low particle velocity and abrasive flow rate were used to conduct the experiments, a higher interaction between incoming and rebounding particles in the region between the nozzle and target was seen since this phenomenon is normally increased at higher impact angles ($\alpha > 45^{\circ}$) as mentioned in previous studies conducted by other investigators [24-26]. The wear damage was characterized by a higher plastic deformation in the central part (primary erosion area). Small pits and craters of up to 50-70 μm in size were observed.

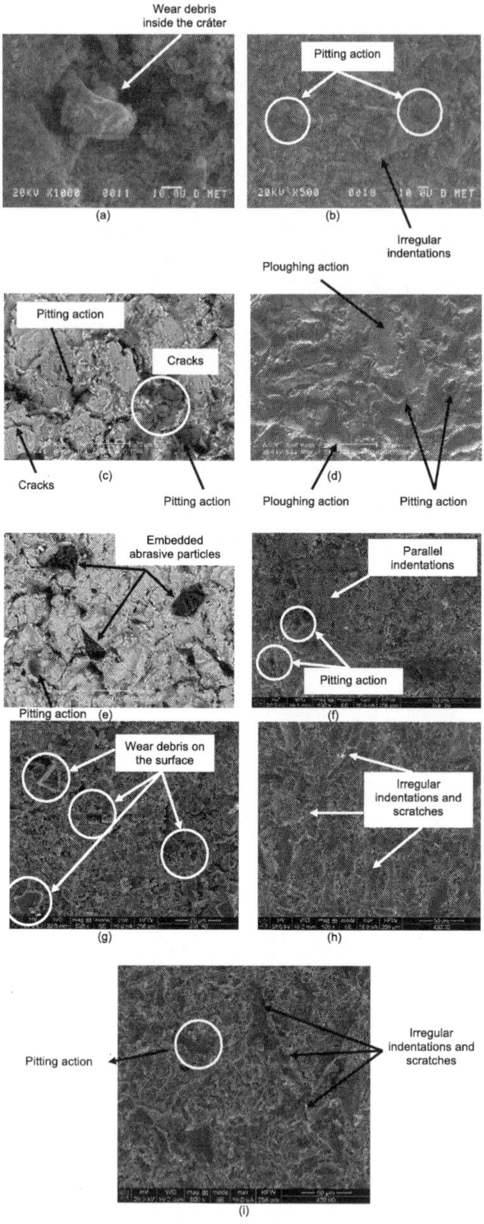

Figure 6. SEM photographs, (a) AISI 4140 at 30°, (b) AISI 4140 at 90°, (c) AISI 1018 at 45°, (d) Stainless steel 304 at 30°, (e) Stainless steel 304 at 90°, (f) AISI 316 at 30°, (g) AISI 316 at 90°, (h) AISI 420 at 30°, (i) AISI 420 at 90°.

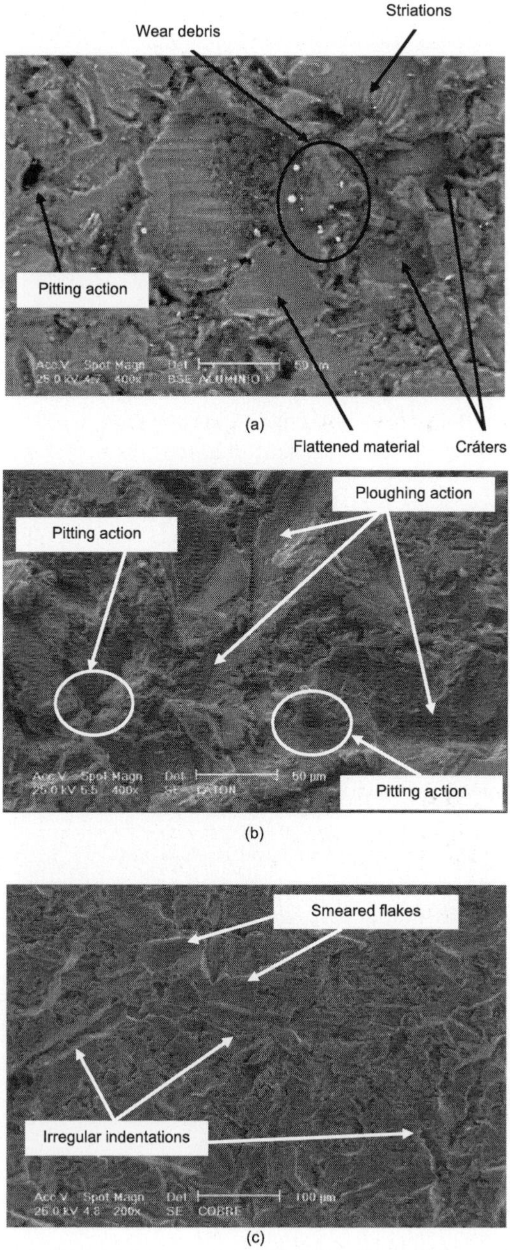

Figure 7. Erosion damage (a) 6061 Aluminium at 90°, (b) Brass at 45°, (c) Copper at 45°

A common occurrence characterized by grooves with material piled up to the sides due to the ploughing action of the particles was observed in brass (Figure 7b). The lips at the sides of the grooves had been flattened due to successive impacts of the erodent particles as exhibited in other studies on metallic materials. In the particular case at 45º (low impact angle), the sliding component played an important role in increasing the mass loss significantly, generating a ploughing action as commonly seen in previous erosion studies. Also, it was possible to see a pitting action which was thought to occur due to abrasive particles that only indented and did not slide on the material surface. Additionally, there is evidence of material separated in form of flakes and elongated parts that were flattened on the copper surface by the subsequent battering of the abrasive particles (Figure 7c) as presented in other erosion experiments [1-4].

Table 3 presents the results obtained of the mass loss at all the incident angles. Additionally, Figure 8 displays a graph of the total erosion rate against the impact angle, where most the erosion rates increases as the angle of impact was decreased. The total erosion rate was obtained by dividing the total mass loss after 10 min of each tested material by the total mass of the erodent hitting the specimen after this time. Most the materials displayed a ductile behaviour as their maximum erosion rate was reached at 30º and 45º and reduced considerably near or at 90º. It is assumed that this initial increase is because of a first group of particles that caused a cutting action on the material surface. In this particular case, the sliding component normally observed at lower impact angles caused severe problems. AISI 304 and 316 exhibited the poorest erosion resistance in comparison with all the tested materials. This behavior was not expected, however the results are very clear. In fact, the maximum erosion rate in these particular cases was reached at 60°, which is not common. Generally, in previous erosion studies on stainless steels, the higher erosion rates are seen at lower impact angles ($\alpha \leq 45°$). It is assumed that the room temperature could be a significant fact to modify the performance of these stainless steels. On the other hand, AISI 420 exhibited a normal behavior, showing its maximum erosion rate at 30°.

Impact angle (α)	AISI 4140	AISI 1018	AISI 304	AISI 316	AISI 420	6061 Al	Brass	Copper
30°	0.0741	0.1372	0.6928	0.5625	0.1011	0.0502	0.1656	0.1119
45°	0.0657	0.1066	0.5803	0.7021	0.0683	0.0471	0.1752	0.1024
60°	0.057	0.0771	1.0875	0.8897	0.0895	0.0455	0.1569	0.0885
90°	0.0536	0.0206	0.7755	0.7398	0.0743	0.0361	0.1103	0.0695

Table 3. Mass loss at different incident angles

The behavior observed in most materials used to conduct this study was as expected because these typically display a maximum erosion rate at lower incident angles and the damage is significantly reduced as the impact angle is increased. It is assumed that the materials used in this study exhibited ductile type behavior. This trend is commonly seen in the graphs used in previous erosion studies, as illustrated in Figure 9 [4].

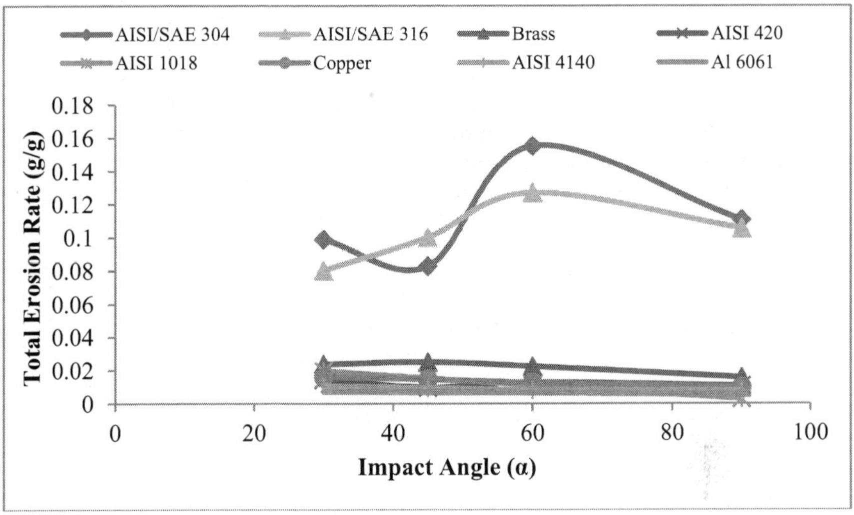

Figure 8. Total erosion rate against impact angle

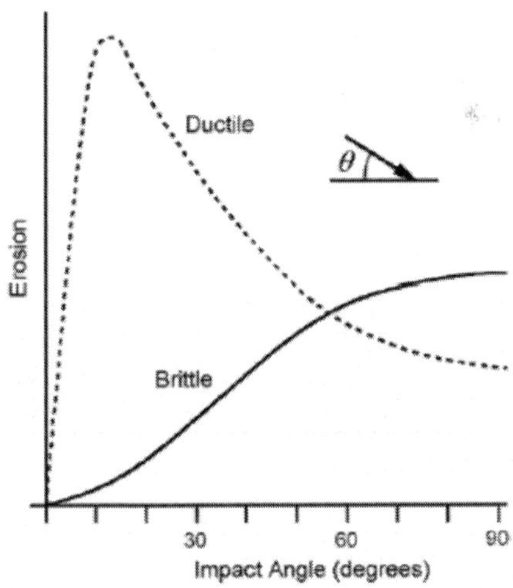

Figure 9. Erosive wear rates for brittle and ductile materials [4].

4. Conclusions

- Aluminium 6061 was the material that exhibited the higher erosion resistance whereas stainless steels 304 and 316 showed the poorer performance against this wear process. It was assumed that the room temperature could have affected the behavior of these materials.
- Most the tested materials exhibited a ductile type behavior due to their maximum erosion rate was reached at lower impact angles (30° and 45°). The erosion rate was considerably decreased at higher incidence angles (60°, 75° and 90°). Stainless steels 304 and 316 had higher erosion damage at 60°.
- Typical wear mechanisms such as ploughing and pitting action, irregular indentations, scratches, craters, embedded abrasive fragments, smeared wear debris on the surfaces and brittle fracture characterized by cracks located at random positions were observed in this particular study.
- The wear scars were characterized by an elliptical shape at 30° and 45°, which is a characteristic feature when the specimens are impacted at low-impact angles ($\alpha \leq 45°$), whereas a nearly circular shape was observed at 60° and 90°.

Author details

Juan R. Laguna-Camacho*
Universidad Veracruzana, Faculty of Electric and Mechanical Engineering, Poza Rica de Hidalgo, Veracruz, México

M. Vite-Torres and E.A. Gallardo-Hernández
2SEPI, ESIME, IPN, Unidad Profesional "Adolfo López Mateos" Tribology Group, Mechanical Engineering Department, México, D.F.

E.E. Vera-Cárdenas
Universidad Politécnica de Pachuca, Pachuca, México

5. References

[1] Venkataraman B. & Sundararajan G., The Solid Particle Erosion of Copper at Very Low Impact Velocities, Wear 1989; 135, 95-108.

[2] Ambrosini L. & Bahadur S., Erosion of AISI 4140 Steel, Wear 1987; 117, 37-48.

[3] Harsha A. P. & Deepak Kumar Bhaskar, Solid Particle Erosion Behavior of Ferrous and Non-ferrous Materials and Correlation of Erosion Data with Erosion Models, Materials and Design 2008; 29, 1745-1754.

[4] Hutchings I. M., Tribology: Friction and Wear of Engineering Materials, London, Edward Arnold; 1992.

---

* Correspondig Author

[5] Morrison C.T. & Scattergood R. O., Erosion of 304 stainless steel, Wear 1986; 111, 1-13.

[6] Levy A. V. & Chik P., The Effects of Erodent Composition and Shape on the Erosion of Steel, Wear 1983; 89, 151-162.

[7] Liebhard M. & Levy A., The Effect of Erodent Particle Characteristics on the Erosion of Metals, Wear 1991; 151, 381-390.

[8] Laguna-Camacho J. R., Development of a prototype for erosion testing using compressed air and abrasive particle flow, Master Thesis, SEPI–ESIME-UZ– IPN, México; 2003.

[9] ASTM standard, G76-95 (1995), Standard practice for conducting erosion tests by solid particle impingement using gas jets, in Annual Book of ASTM Standards, Vol. 03.02, ASTM, Philadelphia, PA, 1995: pp. 321-325.

[10] Vite J., Vite M., Castillo M., Laguna-Camacho J. R., Soto J., Susarrey O., Erosive Wear on Ceramic Materials Obtained from Solid Residuals and Volcanic Ashes, Tribology International 2010; 43, 1943-1950.

[11] Kosel T. H. and Anand K., An Optoelectronic Erodent Particle Velocimeter, in V. Srinivasan and K. Vedula (eds.), Corrosion and particle erosion at high temperature, The minerals, Metals and Materials Society 1989; 349-368.

[12] Shipway P. H. & Hutchings I. M, Influence of Nozzle Roughness on Conditions in a Gas-Blast Erosion Rig, Wear 1993; 162-164, 148-158.

[13] Laguna-Camacho, J. R., A study of erosion and abrasion wear processes caused during food processing, PhD thesis, The University of Sheffield, UK; 2009.

[14] Finnie I., Some Observations on the Erosion of Ductile Metals, Wear 1972; 19, 81-90.

[15] Finnie I., Stevick G. R., Ridgely J. R., The Influence of Impingement Angle on the Erosion of Ductile Metals by Angular Abrasive Particles, Wear 1992; 152, 91-98.

[16] Camacho J., Lewis R., Dwyer-Joyce R. S., Solid Particle Erosion Caused by Rice Grains, Wear 2009; 267, 223-232.

[17] Laguna-Camacho J. R., Cruz-Mendoza L. A., Anzelmetti-Zaragoza J. C., Marquina-Chávez A., Vite-Torres M., Martínez-Trinidad J., Solid Particle Erosion on Coatings Employed to Protect Die Casting Molds, Progress in Organic Coatings 2012; 74, 750-757.

[18] Lapides L., Levy A., The Halo Effect in Jet Impingement Solid Particle Erosion Testing of Ductile Metals, Wear 1980; 58, 301-311.

[19] Hutchings I. M., Winter R. E., Particle Erosion of Ductile Metals: A Mechanism of Material Removal, Wear 1974; 27, 121-128.

[20] Winter R. E., Hutchings I. M., Solid Particle Erosion Studies Using Single Angular Particles, Wear 1974; 29, 181-194.

[21] Tilly G. P., A Two Stage Mechanism of Ductile Erosion, Wear 1973; 23, 87-96.

[22] Tilly G. P., Wendy Sage, The Interaction of Particle and Material Behavior in Erosion Processes, Wear 1970; 16, 447-465.

[23] Bellman R., Levy A., Erosion Mechanisms in Ductile Metals, Wear 1981, 70, 1-27.

[24] Rickerby D. G., Macmillan N. H., The Erosion of Aluminium by Solid Particle Impingement at Normal Incidence, Wear 1980; 60, 369-382.

[25] Shipway P. H., Hutchings I. M., A Method for Optimizing the Particle Flux in Erosion Testing with a Gas-Blast Apparatus, Wear 1994; 174, 169-175.

[26] Anand K., Hovis S. K., Conrad H., Scattergood R. O., Flux Effects in Solid Particle Erosion, Wear 1987; 118, 243-257.

Section 2

Lubrication

Chapter 6

High Speed Rotors on Gas Bearings: Design and Experimental Characterization

G. Belforte, F. Colombo, T. Raparelli, A. Trivella and V. Viktorov

Additional information is available at the end of the chapter

http://dx.doi.org/10.5772/50795

1. Introduction

Gas bearings are employed in a variety of applications from micro systems to large turbomachinery. As they are free from contaminants if supplied with clean air, gas bearings and pneumatic guide-ways are often used in food processing, textile and pharmaceutical industries. The new research works are focused on expanding the applications of gas bearings, in particular at very high speeds. Dental drills for example operate at speeds of over 500 krpm and it seems that a limit for gas bearings without cooling is 700 krpm [1]. Nevertheless in [2] a spindle with 6 mm diameter that operated at 1.2 million rpm is described.

Because of the extremely close manufacturing tolerances that air bearings require and the lack of standard large scale production models, their costs are not at all competitive with those of the rolling bearings in common use. In order to determine whether the initial costs associated with investing in gas bearings will result in savings, each type of technology should be carefully examined. The service life of gas bearings is in fact practically unlimited, since they require almost no maintenance and do not wear.

Many investigations of air bearings have been conducted using experimental, numerical and theoretical approaches with analytical models, e.g. [3-6]. However research is still necessary to improve stiffness, load capacity and stability. At present, research studies potential designs individually to seek the main requirements for a particular application. For dynamic gas bearings, applications are currently limited to those involving low power, though an increasing amount of work is focusing on developing reliable solutions for higher-power uses. Machine tool applications, for example, require a stiffness comparable to those of the rolling bearings in common use; in very high speed applications operational stability is essential. In many cases, parameters such as the number and diameter of supply holes, their arrangement, and supply system geometry come into play. Where rotor stability under low load at very high rotational speed is the prime consideration, designs which bring rotor orbit amplitude down to acceptable levels can be adopted.

The design of gas bearings involves matching the load and stiffness requirements with bearing clearance, orifice type, flow rate and air supply pressure. Numerical calculations can assist bearing design, but their validity must be verified through basic experimental investigations. Therefore at the Mechanical and Aerospace Engineering Department of Politecnico di Torino both experimental and numerical methods were used to design gas bearing spindles and other rotors.

This chapter provides an overview on the design of rotor-gas bearing systems and the experimental activity carried out. For each application developed it is also presented the state of the art that can be found in literature. The models developed to simulate the rotor-bearings systems are described in a separate paragraph.

Four prototypes of high speed spindles were designed using gas bearings: a completely pneumatic spindle, an electro-spindle designed for machine tools, a rotor for textile applications and a mesoscopic spindle devoted to high precision machining of micro-parts at very high speeds.

2. The pneumatic spindle

In high speed machining there are some applications for drilling, milling, and grinding, in which gas bearings are used to support the spindle [7,8]. The spindle technology in ultra-precision turning and grinding is nowadays an integration of the motor, spindle shaft and the bearings. In general these spindles have diameters smaller than 20 mm and it is difficult to find an application with a pneumatic spindle of greater diameter. In reference [9] a prototype for woodworking with spindle diameter 60 mm is described.

The prototype developed at Politecnico di Torino is capable of achieving 100000 revolutions per minute and operates at an air supply gauge pressure of 0.4-0.6 MPa. It was designed with the purpose of obtaining high load capacity and stiffness on bearings, so the spindle diameter is greater than spindles designed to achieve 200000 rpm.

The spindle is shown in Figure 1, which illustrates how the housing (4) is constrained to the base through flange (5) and journal (6). Radial support is provided by bushings (7) and (8), while axial thrust is opposed by disks (9)-(11).

The housing is made of 18 Ni Cr Mo 5 steel, while the rotor (Figure 2) (mass 7 kg, diameter 50 mm, length 459 mm) is made of 88 Mn V 8 Ku tool steel quenched and tempered to a hardness of 60 HRC. The rotor was also aged in liquid nitrogen for 5 h and dynamically balanced to a grade better than ISO quality grade G-2.5. The nose (12) to which loads are applied is secured to one end of the rotor, while the driving turbine (13) is integral with the other end. Bushings are made of the same material as the housing and have an axial length of 100 mm.

The bearings were designed to maximize the stiffness because of the importance of this parameter during cutting operations. They are provided with four circumferential sets of four 0.25±0.01 mm diameter radial holes, drilled in brass inserts as shown in Figure 3.

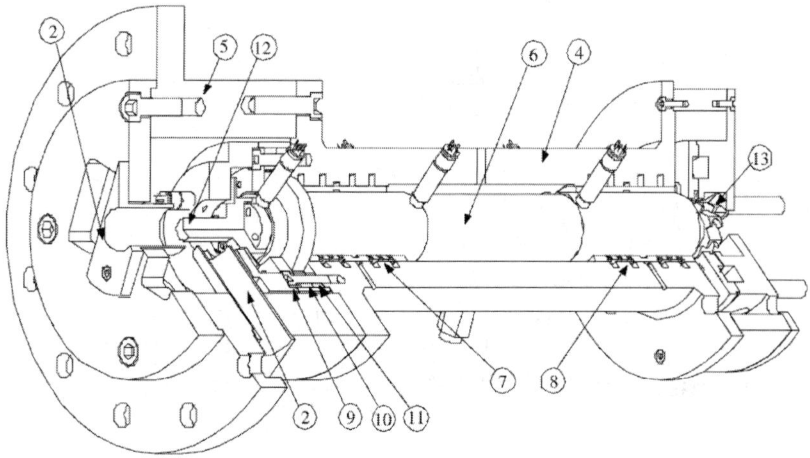

Figure 1. Section of the pneumatic spindle

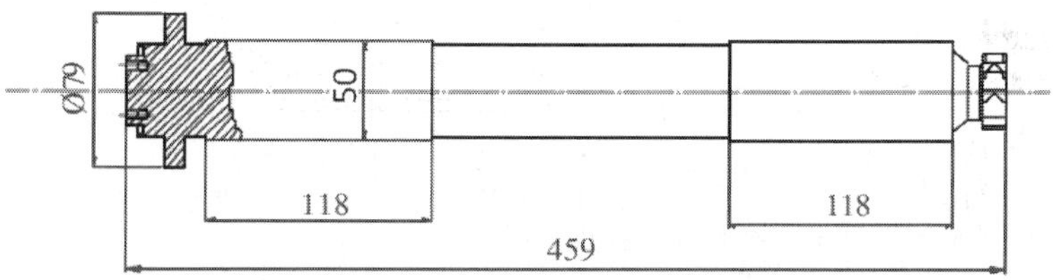

Figure 2. Rotor

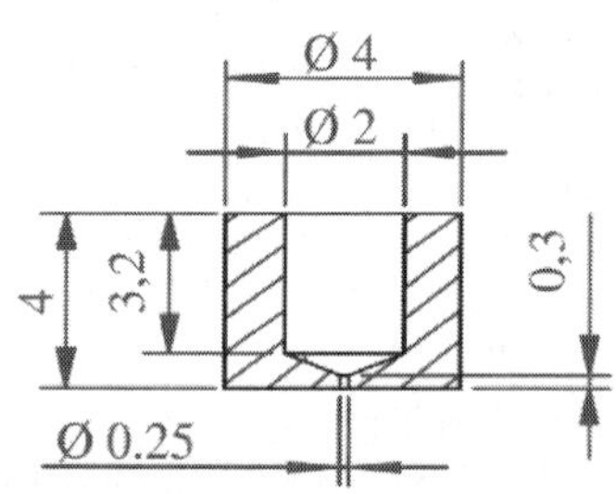

Figure 3. Brass insert with the supply hole

The axial thrust (Figure 4) is controlled by two disks (9) and (11) facing the flange on the journal. These disks are separated by a ring (10) whose thickness determines the size of the air gap. Supply air is delivered from an axial hole in the housing, is distributed through a circumferential slot, and then crosses a series of axial and radial channels machined in the disks to reach 0.25±0.01 mm diameter axial nozzles (14) and (15), which are also machined in inserts. Both the bushings and the disks were surface hardened and machined to produce a surface roughness of 0.2 and 0.4 μm, respectively at the air gaps.

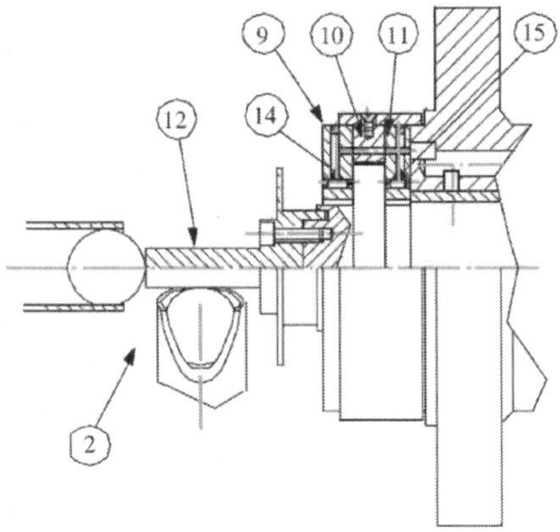

Figure 4. Enlargement of the thrust bearing and the nose

Radial and axial forces are applied to nose (12) by means of load devices (2). These devices are made of a hollow cylinder containing a calibrated sphere with a diametral clearance of 40 μm. When the cylinder chamber is supplied, the sphere is pushed against the nose and at the same time supported, so that it can rotate against the nose without sliding. Radial and axial forces can thus be transmitted to the rotor even when the latter is in motion.

Supply air for the turbine (Figure 5) crosses pre-distributor (16) in the axial direction to reach annular chamber (17), from which distributor (18) leads to eight tangential channels. Air is exhausted after actuating the turbine. Open loop speed control is accomplished by establishing turbine supply pressure.

Bearing supply is separate from turbine supply. Should the air supply fail, a reservoir enables the bearings to operate during rotor deceleration, thus preventing the rotor from seizing on the bushings or disks. Supply lines are provided with two air filtration units featuring borosilicate glass microfiber cartridges whose filtration efficiency is 93 and 99.99% respectively with 0.1 μm diameter particles. For the bearing supply line, an activated carbon coalescent filter was added to eliminate any oil vapors.

The test bench uses five capacitive displacement transducers with 0.1μm resolution, 500 μm full scale reading and 6 kHz passband. One of the transducers is used axially to measure the relative position of the rotor and thrust disks. The other four are installed radially on two different planes at right angles to the rotational axis. Rotor displacement in the bushing can thus be measured in both plane directions. The signals from these sensors are amplified by appropriate charge preamplifiers and then conditioned in a module containing a single oscillator and a demodulator for each channel. The sensitivity of these sensors is constant within the 0-10 V linearity range.

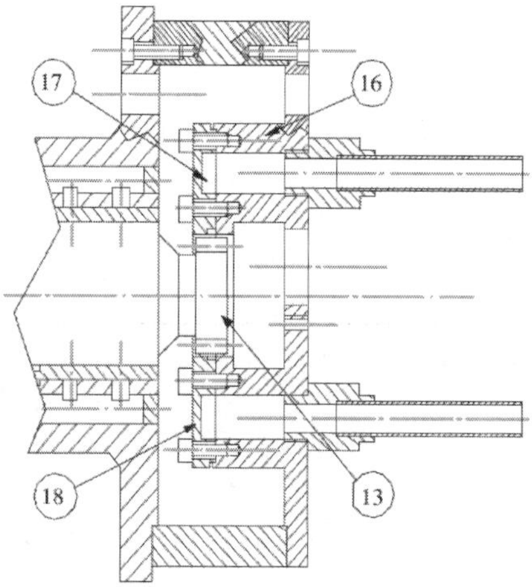

Figure 5. Enlargement of the driving turbine

To determine the journal rotational speed, an optical tachometer provided with an emitter and receiver is used, with a digital counter. Several thermocouples are also used to measure the temperature of the outer housing surface and of the air issuing from the bearing exhaust ports.

Dimensional checks were carried out to know with good precision the air gap of the bearings and the diameters of the holes. The mean inside diameter of the bushing and the mean external diameter of the rotor were measured with a precision height gage (Mitutoyo Linear Height). The mean radial air gap was calculated as the difference between the radius of the bushing and the radius of the rotor. The axial mean air gap between thrust flange and disks is calculated as the difference between measured ring and flange mean thicknesses. Results are compared with the nominal air gaps in Table 1.

	Radial air gap (μm)	Axial air gap (μm)
exp. value	26	18
nominal value	25	20

Table 1. Measured values of the air gaps

The diameters of the holes that supply the bearings were checked with an optical fiber camera with 50x and 100x magnifying lenses. Table 2 shows the measured mean diameters of the holes, with the indication of the frequency. These holes were produced using microdrills.

Diameter (mm)	Frequency
0.24	25
0.245	7
0.25	16

Table 2. Measured diameters of the supply holes

Tests were carried out to determine the bearing stiffness with the rotor stationary. The radial stiffness measured in correspondence of the nose at 120 mm from the front side of the bearing is 18 N/μm at 0.6 MPa supply gauge pressure. The axial stiffness is 27 N/μm at the same supply pressure.

Figure 6 shows the thermal transient at 40 krpm for spindle internal and external temperature measurements. The internal temperature is close to that of the air issuing from the exhaust ports. This is in accordance with the results indicated in the literature, see e.g. [10].

Rotor orbits at the two radial bushings were recorded at speeds up to 50 krpm, although tests have gone up to 80 krpm.

Figure 7 shows an example of orbits at 45 krpm, both in forward precession. Sensors 1 and 2 are for the bushing on the turbine side, while 3 and 4 are for the bushing on the motor side. These orbits, which were measured with zero radial and axial loads, are synchronous and stable. As signal frequency analysis indicated that no peak appears at a frequency of around half the rotation frequency, unstable whirling does not occur.

The centrifugal forces effect has been taken into account during the rotor designing. The radial deformation of the rotor far from its flange is visible in Figure 8.

The approaching of the external surface of the rotor to the sensors has also been considered in order to plot the orbits.

By means of a finite element code a circumferential groove was designed in proximity of the rotor flange in order to compensate the deformation due to the centrifugal force of the flange.

In Figure 9 the calculated deformation with the circular groove (depth 0.5 mm, length 10 mm) is visible. The deformation is enlarged with respect to the rotor profile. Also thermal effects on the relative distance between rotor and sensors, mounted on the housing, have been taken into account in order to individuate the centre of the orbit.

3. The electro-spindle

In literature can be found examples of air bearing electro-spindles for high speed and high precision applications, see for example references [11-13]. The electro-spindle developed at Politecnico di Torino [14], shown in Figure 10, is composed of a rotor of 7 kg mass, 50 mm diameter and 479 mm length. It is supported by air bearings and accelerated by means of an asynchronous motor mounted on one end of the spindle. On the opposite end of the rotor a clamping tool is mounted.

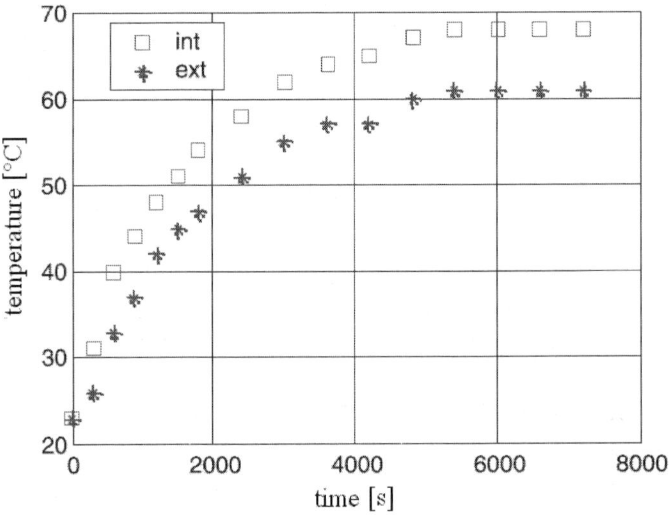

Figure 6. Thermal transient

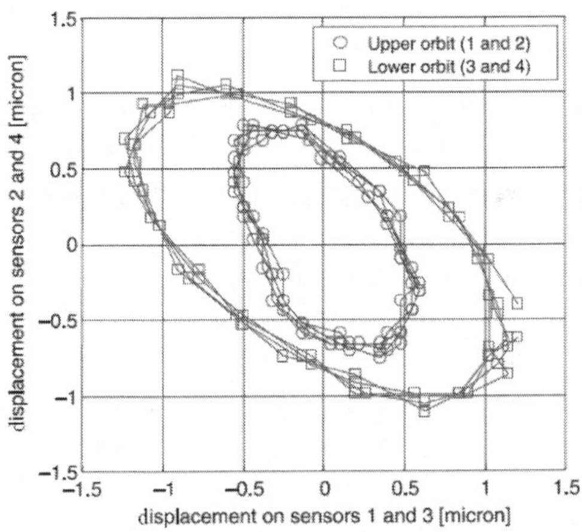

Figure 7. Rotor orbits; ω=45 krpm

Figure 11 shows a section of the electro-spindle with carter (1), rotor (2), two bushings (3) and double thrust bearing (4). Motor (5) is of the two-pole squirrel-cage type controlled by an inverter. Speed range is up to 75 krpm and power is 2.5 kW. A clamping tool designed for high-speed is screwed onto the left end of the rotor. By mounting a tool on the spindle it is possible to test the dynamic behaviour of the rotor also during the machining process.

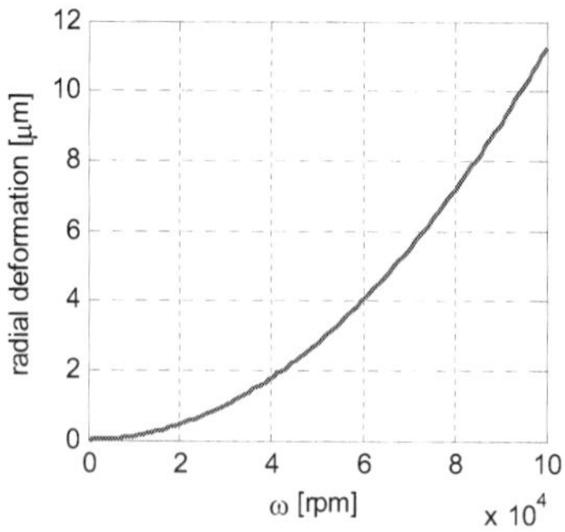

Figure 8. Centrifugal expansion of the rotor in correspondence of the bushings (diameter 50 mm)

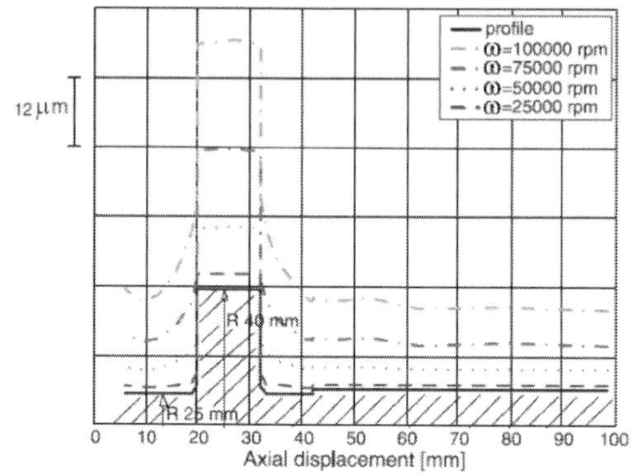

Figure 9. Rotor deformation due to centrifugal force in correspondence of the flange at different rotational speeds

The radial bearings feature cylindrical barrels. Each barrel has four sets of supply ports diameter 0.25±0.01 mm arranged 90 degrees apart. The thrust bearing, similar to the one described in [15], is composed of two disks facing the flange on the journal. Each disk has 8 axial nozzles dia. 0.2±0.01 mm positioned on the mean diameter.

The system is provided with a closed cooling circuit that controls the temperature of the motor and of the discharge air. Without refrigeration and with ambient temperature 298 °K, at 60000 rpm the temperature would reach 383 °K after two hours due to power losses on bearings.

An optical tachometer facing the rotor was provided to measure rotational speed. Four capacitance displacement transducers were inserted radially and at right angles in the carter facing the rotor to measure dynamic runout. Two were positioned on motor side, the other two on thrust bearing side.

Clearances were measured moving the rotor axially and radially until contact is made. It was found that axial and radial clearances are about 15 µm and 20 µm respectively.

In order to measure radial and axial stiffness of the tool, the electro-spindle is mounted on a test rig designed for the purpose with proper load devices. Radial forces were measured at different supply pressures (Figure 12) at ω=0. Figure 13 shows the axial load capacity readings for 0.3, 0.5 and 0.7 MPa supply absolute pressure. The load device was used for positive displacements and weights were applied to obtain the curve with negative displacements.

The rotor orbits depicted in Figure 14 were measured at the same supply pressure. Due to the rotor centrifugal expansion these orbits appear not to be centered in the bushings because the relative rotor-sensor distance decreases. The spindle was tested up to 53000 rpm and the tests were stopped because of the high rotor vibration. The permissible residual imbalance should be diminished in order to allow tests at higher speeds. Anyway the whirl instability did not occur and the imbalance response was only synchronous.

Figure 10. Photo of the electro-spindle

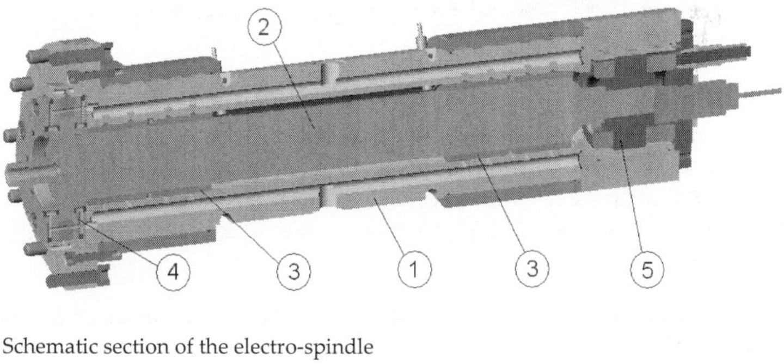

Figure 11. Schematic section of the electro-spindle

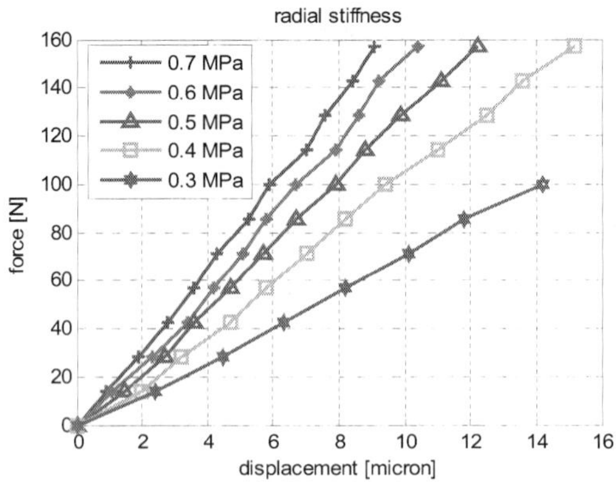

Figure 12. Radial force on the tool (measured at 120 mm from the front side of the bearing) versus radial displacement at different bearing supply absolute pressures

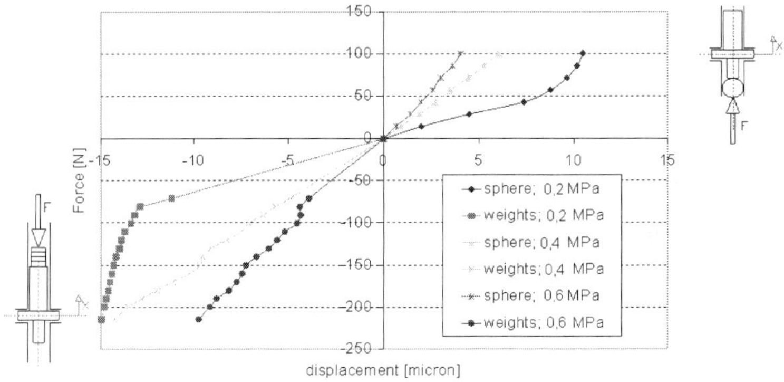

Figure 13. Diagram of axial force on tool versus displacement at different bearing supply gauge pressures

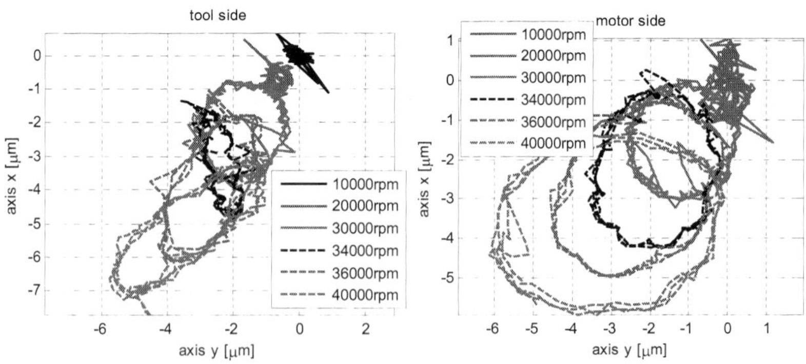

Figure 14. Rotor orbits in correspondence of a bushing due to the residual unbalance; supply gauge pressure 0.6 MPa

The electro-spindle was also tested in dynamic conditions during machining with high speed milling cutters of diameters in the range 1 to 6 mm. The system depicted in Figure 15, mounted below the electro-spindle, provides the advance along axis x of the material under milling. The material under machining was a block of rapid prototyping resin, advanced by means of a motorized slide. The tests were made up to 40000 rpm with feed speeds from 1 to 10 mm/s and chip thickness 1 mm.

Figure 15. Motorized slide used for the dynamic tests

4. The textile rotor with damping supports

Gas bearings suffer from instability problems at high speed. A method to increase the stability threshold (the speed at which the unstable whirl occurs) is to increase the damping of the rotor-bearings system by introducing external damping supports [16]. A design guideline for the selection of the support parameters that insure stability in an aerodynamic journal bearing with damped and flexible support is given in paper [2].

The prototype described in this paragraph was designed with the priority of increasing the stability at high speeds [17]. The method adopted for this purpose was the use of rubber O-rings.

The prototype consists on a rotor (1) made of hardened 32CrMo4 steel with mass 0.96 kg, diameter 37 mm and length 160 mm. The rotor is supported by a radial air bearing mounted on rubber O-rings and an axial thrust bearing (Figure 16). It was designed to rotate in stable conditions up to 150 krpm. At one end of the rotor an air turbine (2) was machined and at the other end a nose (3) was screwed to the rotor. The housing (4) is fixed to the base and has four circumferential slots in which the O-rings are inserted. The bushing (5) incorporates the rubber rings and has four sets of supply nozzles (diameter 0.2±0.01 mm) fabricated by EDM. The total length of the bearings is 57 mm. In the middle plane of the bushing a discharge slot (6) is vented by a radial hole in the housing (see Figure 17). A central annular discharge chamber separates the radial bearings.

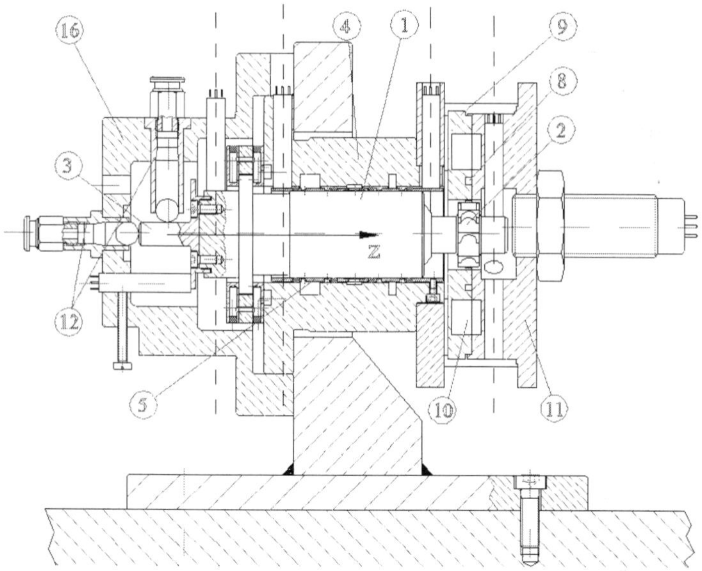

Figure 16. Test bench of the floating bushing

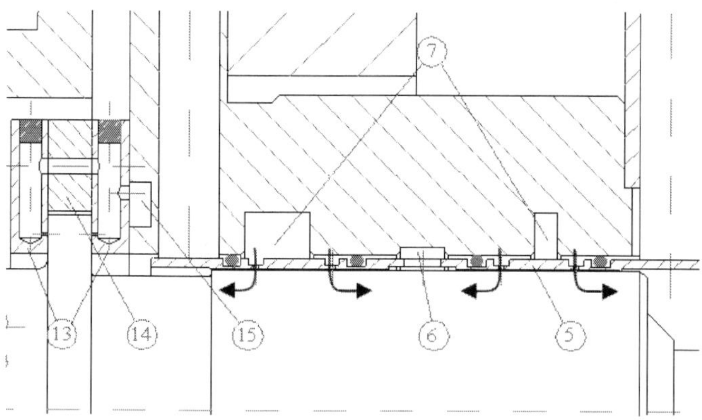

Figure 17. Enlargement of the floating bushing

The air from supply slots (7) flows to the radial clearance through the nozzles to reach the vent centrally in the discharge slot and laterally. The purpose of the O-rings, besides providing a seal between supply slots and discharge chamber, is to introduce damping in the rotor-bearing system. The turbine is driven by tangential jets discharged through 8 nozzles (8) machined on distributor (9). Annular chamber (10) connected to the nozzles is supplied through an axial hole on pre-distributor (11). Air is exhausted after actuating the turbine. Open loop speed control is maintained by setting the turbine supply pressure. The rotational speed was measured by an optical tachometer consisting of an emitter and a receiver facing the rotor at the turbine side. A retro-reflector stuck to a portion of the rotor face reflects emitted signal once per revolution.

Radial and axial forces are applied to the nose by means of loading systems (12) similar to the ones previously described. Eight capacitive displacement transducers are inserted radially in the housing, the pre-distributor and cover (16) to sense the rotor and bushing positions. An axial transducer can be inserted near the nose to monitor the axial position of the rotor with respect to the thrust bearing.

The O-Rings have 41 mm inside diameter and 70 Shore hardness. The three materials used for testing are NBR (Butadiene Acrylonitrile), Viton® (Fluorinated Hydrocarbon) and Silicone (Polysiloxane).

Accurate dimensional checks were carried out to evaluate axial and radial clearances, supply holes diameter and O-ring interference. The total diametral gap between them was found to be 35±2 μm. The difference between the thickness of central ring (14) and rotor flange was 19±2 μm, giving an axial clearance of approximately 9.5 μm.

To measure the diameter of the nozzles supplying the bearings an optical fibre camera with 200X magnifying lens was used. The measurements were accurate and repeatable, thus proving the superiority of EDM technology over micro-drilling. Figure 18 shows a sample photographic record at 200X magnification.

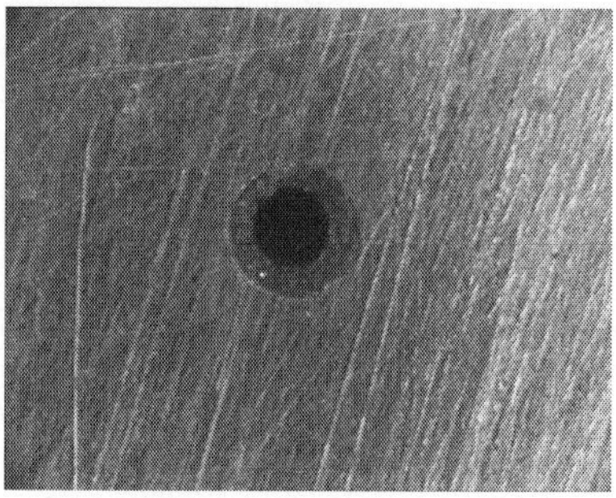

Figure 18. Supply hole magnification (200X)

The supply hole diameter, after fixing the radial clearance, is selected on the basis of numerical investigation conducted to simulate the dynamic behaviour of the system. The mathematical model used for this purpose is described in a separate paragraph at the end of this chapter.

O-ring grooves in the housing have a medium diameter of 43.5 mm, while external diameter of the bushing is 41 mm. The cross section diameter d of the rings was determined by a shadow comparator.

Table 3 lists the interferences on the O-rings calculated using the equation

$$int\% = \left(d - \frac{D_i - D_e}{2}\right)\frac{100}{d}$$

where D_i and D_e are the inside diameter of the grooves machined in the housing and the external diameter of the bushing respectively. With 0.6 MPa pressure differential the sealing function of the rubber rings between chambers 7 and 6 was realized with interference about 10% or more.

In Table 4 the measures of the inner diameter d and the cross-section diameter d_c are shown.

	Cross section diameter d [mm]	Interference
NBR-Silicone	1.78±0.01	30%
Viton	1.73±0.01	28%

Table 3. Interference values

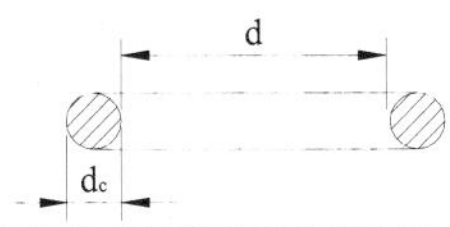

	d (mm)	d_c (mm)
NBR	41	1.80
Viton	41	1.83
Silicone	41	1.83

Table 4. Dimensions of the O-rings

4.1. Measured rubber dynamic stiffness

The dynamic stiffness of rubber O-rings is measured in order to introduce into the model the stiffness and the viscous equivalent damping. These parameters depend on the vibration frequency and also on the radial displacement imposed. Tests were made under different conditions, varying the diametral interference on the O-ring and the displacement amplitude x_0 imposed, in the frequency range 300÷800 Hz. In Figure 19 is visible the scheme of the test rig, in which the cylinder is fixed and the casing is mounted on the shaker plate. The force amplitude F_0 is measured by the load cell mounted between two fixed parts.

Measurements were made at different radial amplitudes. Increasing x_0 both stiffness and damping decrease. The results visible in Figure 20 were obtained with x_0=25 μm and a diametral interference of 11%, that are similar to that occur in the air bearing test bench. The results obtained with Silicone are not reported because the FRF of transfer function F/x was very noisy.

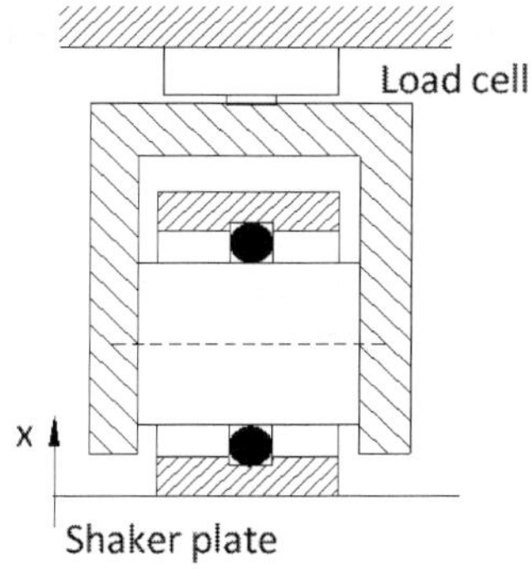

Figure 19. Scheme of the O-ring test rig

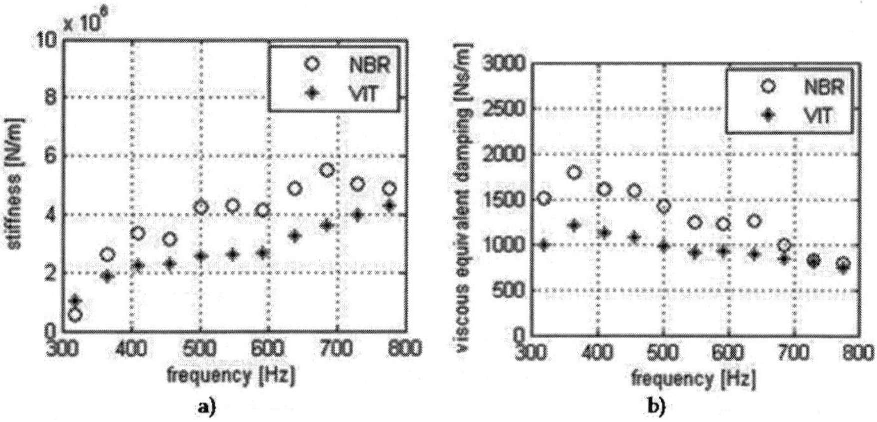

Figure 20. O-ring radial stiffness (a) and damping (b)

4.2. Stability

The response to a rotor radial step jump displacement of 1 μm from coaxial position is calculated. As a first approximation, average values of O-ring stiffness k_{OR} and damping c_{OR} are considered, neglecting the dependence on the frequency.

The parameters introduced in the model are shown in Table 5. L_1 and L_2 are the axial lengths of the two radial bearings.

m_{rot}=0,97 kg	h_0=17 μm	L_2=23 mm	T^0=293 K
m_b=0,1 kg	d_s=0,2 mm	μ=1,81e-5 Ns/m^2	k_{OR}=4·10^6 N/m
R=18.5 mm	L_1=25 mm	R^0=287 J/kgK	c_{OR}=1·10^3 Ns/m

Table 5. Input values of the model

In Figure 21 the theoretical supply pressure values in correspondence to the stability threshold are plotted vs. the rotational speed for the cases of fixed bearing and bearing mounted on O-rings. Each curve divides the plane into two regions: the upper one relative to a stable behaviour of the rotor-bearing system, the lower one relative to an unstable behaviour. In the first case, as a result of an initial step jump displacement of the rotor, the system evolves to the centred position (punctual stability); in the second case the rotor trajectory is an open spiral and causes the contact between the rotor and the bushing. In correspondence to the threshold curves the system evolves to a condition of orbital stability. The stabilizing effect of the rubber rings is evident because the pressure that guarantees the stability is lower.

In Figure 22 the simulated values are compared with the experimental ones, relative to three kinds of rubber: NBR, Viton and Silicone. There is good agreement between the experimental and the simulated stability threshold also if the experimental data are influenced by the rotor imbalance and in calculations the effect of imbalance is neglected (the rotor was dynamically balanced to a grade better than ISO quality grade G-2.5).

The whirling frequency◎ ν increases with the rotational speed, see Figure 23. It is interesting to observe that the whirling ratio $\gamma = \nu/\omega$ at the stability threshold (Figure 24) decreases with the rotational speed.

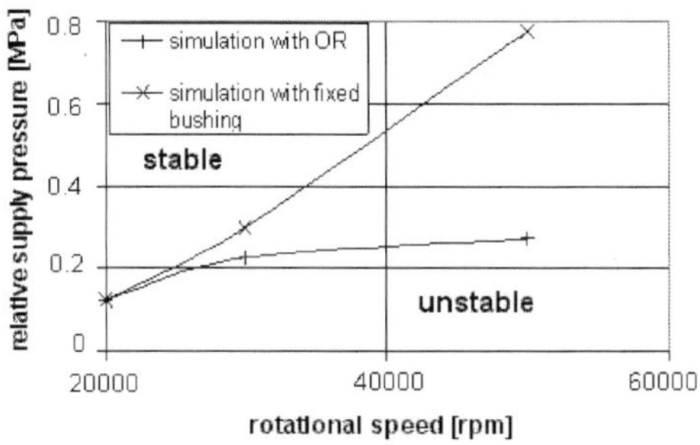

Figure 21. Theoretical results with fixed bearing and bearing mounted on OR

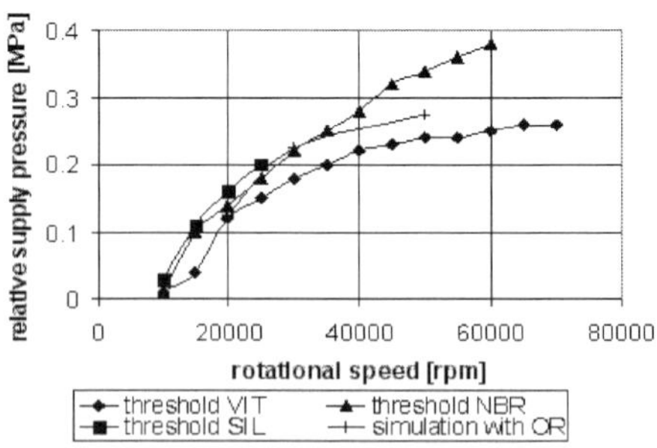

Figure 22. Comparison between experimental and simulated threshold stability

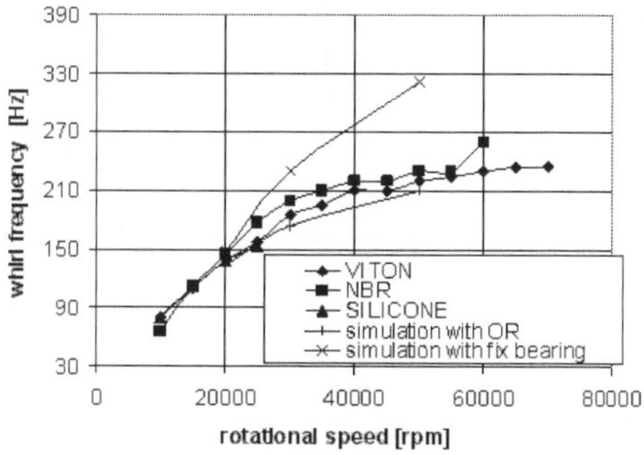

Figure 23. Comparison between experimental and simulated whirling frequency ν

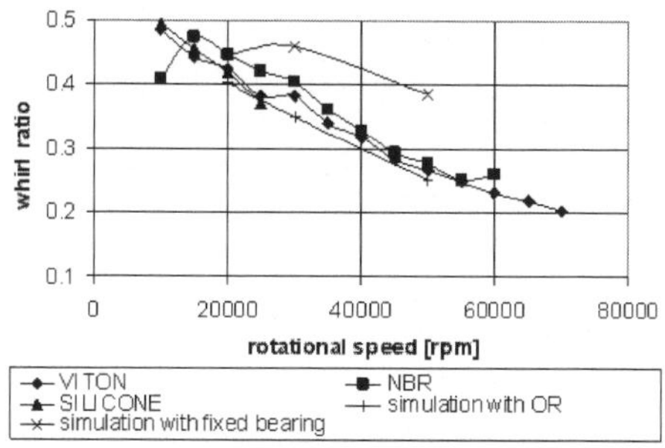

Figure 24. Comparison between experimental and simulated whirling ratio γ

It is possible to approach this threshold by decreasing the supply pressure or by increasing the rotational speed. Both possibilities are treated: Figures 25 and 26 show the change of the orbit amplitude vs these parameters. The increase in amplitude near stability threshold is sudden and considerable in both cases.

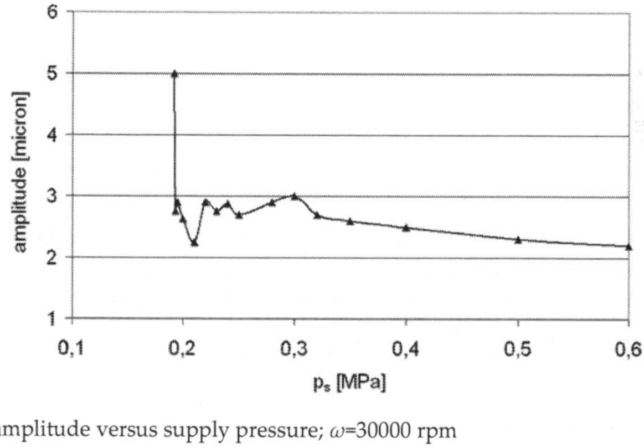

Figure 25. Orbit amplitude versus supply pressure; ω=30000 rpm

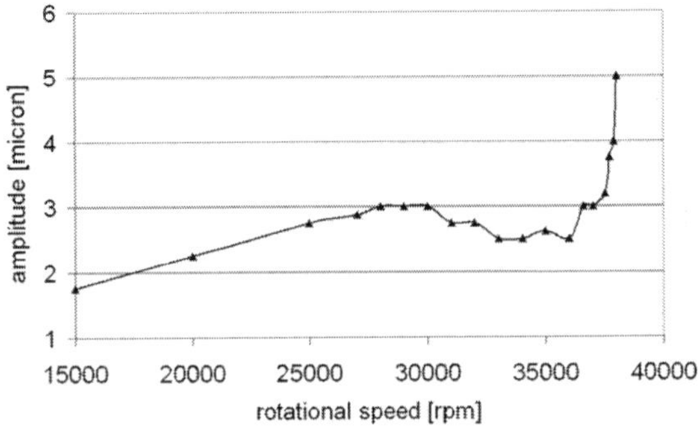

Figure 26. Orbit amplitude versus rotational speed; p_s=0.22 Mpa

Figure 27 shows the change in orbit shape with decreasing the supply pressure. Whirl motion is conical for any supply pressure at the stability threshold. Frequency spectra for turbine displacement in the two conditions are visible in Figure 28.

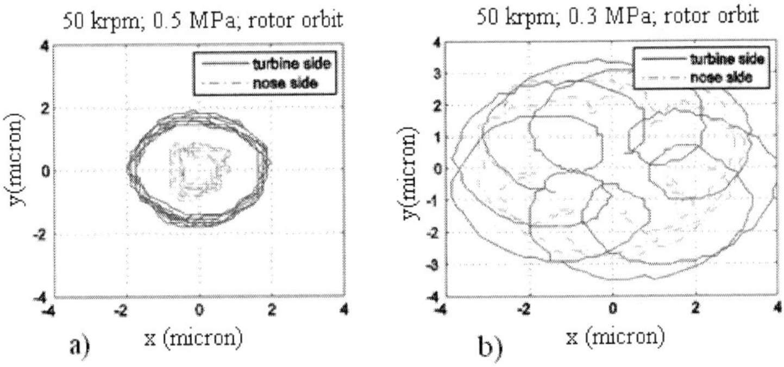

Figure 27. Rotor orbits in stable condition (a) and at stability threshold (b); ω=50000 rpm

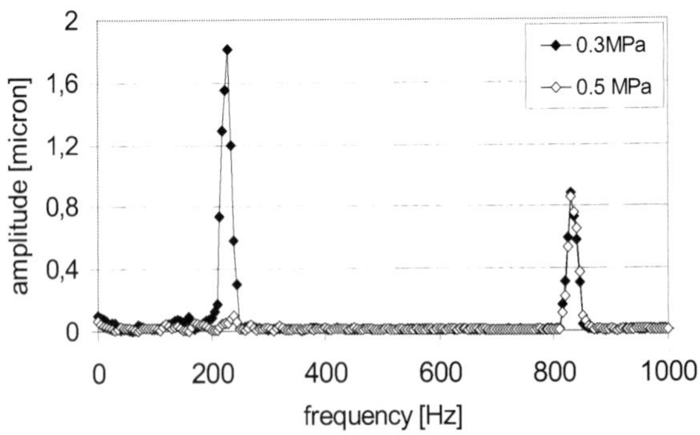

Figure 28. Frequency spectra of rotor displacement; ω=50 krpm

5. The mesoscopic spindle

Another method to increase the bearing stability is to modify the film geometry from the circular journal bearing profile. Non-circular journal bearings can assume various geometries: elliptical [20-23] offset halves [24] and three-lobe configuration [25,26] are the most common geometries. Paper [27] shows a comparative analysis of three types of hydrodynamic journal bearing configurations namely, circular, axial groove, and offset-halves.

There is an extensive literature about the study of the dynamic stability of hydrodynamic journal bearings with non-circular profile, but very few papers consider gas journal bearings of this type. The wave bearing with compressible lubricants was introduced in the early 1990's [28,29].

In the present paragraph the design of the elliptical and multi-lobes gas bearings for a ultra-high speed spindle is described [30].

The bearings were designed to have a stable regime of rotation up to 500 krpm with acceptable stiffness and load characteristics. A computerized design was used for optimization of the rotor-bearing characteristics. The bearing clearance was represented by expression

$$h = h_0\left(1 + \frac{c_{form}}{2}(cos(n\vartheta) - 1)\right)$$

where c_{form} is the profile form factor, h_0 is the maximum clearance and n is the number of lobes of the profile.

The static and dynamic performances were numerically analyzed for two pairs of radial externally pressurized gas bearings. Conical and cylindrical whirl modes were considered. From numerical simulations for a 10 mm diameter rotor, bearing clearances non less than 5 μm and supply pressure 0.6 MPa the following results were obtained:

- the maximum rotor speed obtained with circular bearing (clearance 5 μm) with 4 supply orifices of 0.1 mm diameter in circumferential direction was 150 krpm, while with 32 supply orifices of 0.2 mm diameter was 250 krpm;
- with elliptical bearing profile the maximum rotor speed obtained with stable operation was 500 krpm for bearings with 4 supply orifices of 0.2 mm diameter in circumferential direction;
- the rotor with the multi-lobe bearings were less stable in comparison with the rotor with elliptical bearings;
- the positioning of supply orifices at 45° with respect to the principal axes of the elliptic profile improved bearing characteristics.

The final bearing geometry is defined by the parameters summarized in Table 6. Each elliptical journal bearing presents two rows of 4 supply orifices positioned at 45° with respect to the principal axes.

Maximum clearance h_0, µm	15
Rotor diameter, mm	10
Supply orifice diameter, mm	0.2
Number of supply orifices for each bearing	8
Number of bearings	4
Profile form factor c_{form}	0.7
Number of profile lobes n	2

Table 6. Final bearing parameters

Figure 29 shows the prototype of ultra-high speed spindle. The rotor, of mass 0.07 kg, is supported by two pairs radial elliptical bearings and a double thrust bearing. The calculated radial stiffness on the rotor end is 3 N/µm and the air consumption is $3.65 \cdot 10^{-4}$ kg/s.

The axial and the radial stiffness of the bearings were measured with test benches realized at the purpose (Figure 30). The axial and radial displacement of the rotor due to an imposed load was measured by laser beams. The axial stiffness of the thrust supplied at 0.6 MPa is 2.8 N/µm, while the radial stiffness is 1 N/µm. This value can be increased with a better dimensional control of the bearings internal profile.

Figure 29. Ultra-high speed spindle prototype

By means of start-up (acceleration) and coast down (deceleration) tests on the spindle the bearing friction torque was estimated as a function of the speed up to 150000 rpm. The deceleration tests from different rotational speeds are depicted in Figure 31. The friction torque was found to be proportional to the rotational speed with the rate of 10^{-4} Nm every 10000 rpm. The dynamic runout of the shaft was measured by means of laser beams at different rotational speeds in correspondence of the nose.

The unbalance response was synchronous and unstable whirl was not encountered. In Figure 32 the waterfall diagram, obtained with the FFT of the shaft radial vibration, is shown. There is a critical speed at 34000 rpm, to which corresponds a maximum spindle runout of ±9 µm.

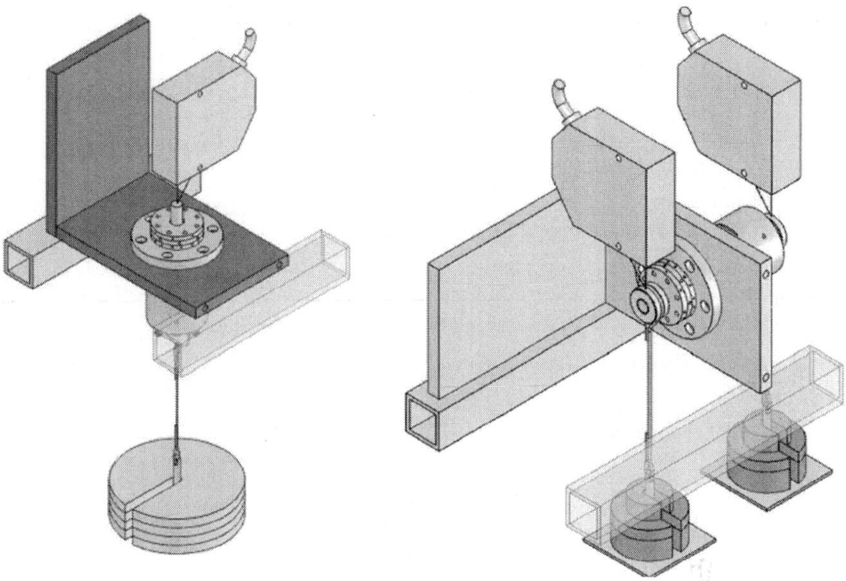

Figure 30. Test benches realized to measure the radial and axial bearing stiffness

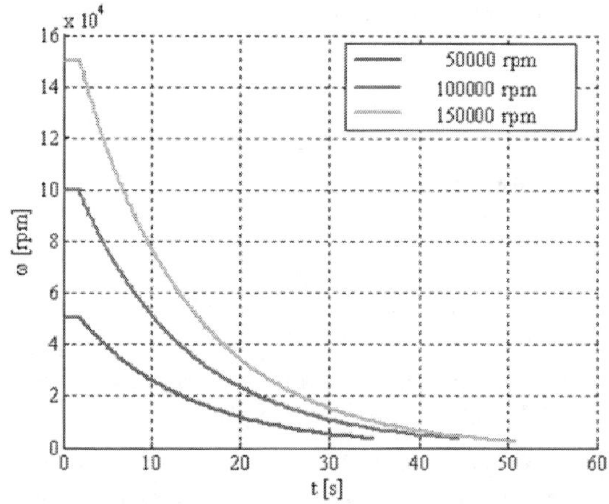

Figure 31. Test benches realized to measure the radial and axial bearing stiffness; supply gauge pressure 0.6 MPa

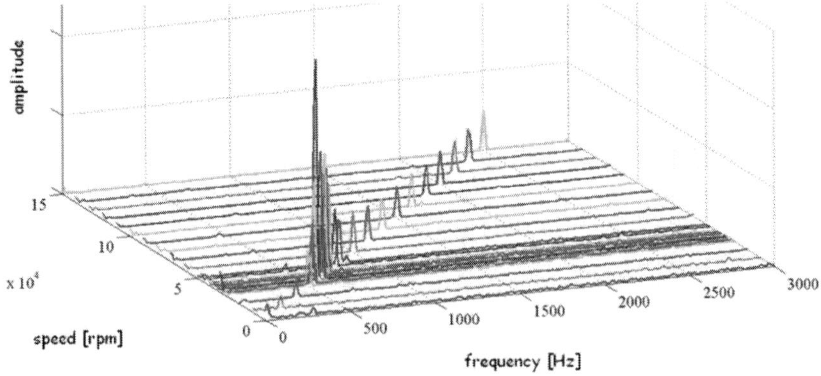

Figure 32. Waterfall diagram of the rotor unbalance response

6. Mathematical model

The complete Reynolds equation for compressible fluid film is numerically solved together with the equations of motion of the rotor considered rigid, see reference [18].

The momentum equations for the isothermal gas lubricated films are:

$$\frac{p}{R^0T^0}\left(\frac{\partial u}{\partial t}+u\frac{\partial u}{\partial z}+v\frac{\partial u}{R\partial\vartheta}\right)+\frac{\partial p}{\partial z}=-\mu\frac{12u}{h^2} \tag{1}$$

$$\frac{p}{R^0T^0}\left(\frac{\partial v}{\partial t}+u\frac{\partial v}{\partial z}+v\frac{\partial v}{R\partial\vartheta}\right)+\frac{\partial p}{R\partial\vartheta}=\mu\frac{6(R\omega-2v)}{h^2} \tag{2}$$

where u and v are the mean velocity components in z- and θ-direction (Figure 33).

They are solved together with the continuity equation

$$\frac{\partial(phu)}{\partial z}+\frac{\partial(phv)}{R\partial\vartheta}+\frac{\partial(ph)}{\partial t}-R^0T^0q=0 \tag{3}$$

where q is the inlet mass flow rate per unit surface defined by

$$q=\frac{G}{\Delta zR\Delta\vartheta} \tag{4}$$

For low modified Reynolds numbers ($Re^{*}=\rho\omega h_0^2/\mu<1$) inertial terms are negligible. The equation resulting from (1-3) is simplified into the following

$$\frac{\partial}{\partial z}\left(ph^3\frac{\partial p}{\partial z}\right)+\frac{\partial}{R\partial\vartheta}\left(ph^3\frac{\partial p}{R\partial\vartheta}\right)+12\mu R^0T^0q=6\mu\omega\frac{\partial(ph)}{\partial\vartheta}+12\mu\frac{\partial(ph)}{\partial t} \tag{5}$$

Film thickness h is given by

$$h(\vartheta,z)=h_0-e_x(z)cos\vartheta-e_y(z)sin\vartheta \tag{6}$$

The rotor eccentricities e_x and e_y are related to the journal degrees of freedom by

$$e_x(z) = x_G + (z - z_G)\vartheta_y;\ e_y(z) = y_G - (z - z_G)\vartheta_x \tag{7}$$

Concerning the thrust bearing, the Reynolds equation is

$$r^2 \frac{\partial}{\partial r}\left(ph^3 \frac{\partial p}{\partial r}\right) + \frac{\partial}{\partial \vartheta}\left(ph^3 \frac{\partial p}{\partial \vartheta}\right) + 12\mu R^0 T^0 r^2 q = 6\mu\omega r^2 \frac{\partial (ph)}{\partial \vartheta} + 12\mu r^2 \frac{\partial (ph)}{\partial t} \tag{8}$$

The boundary conditions at the discharge slots are $p=p_a$, while in correspondence of the supply ports the downstream pressure level is calculated considering orifice resistance.

The input resistance is expressed on the basis of ISO formula for flow rate through an orifice (ISO, 1989).

$$G = \begin{cases} c_s k_T \rho_N p_s & if\ 0 < \frac{p}{p_s} < b \\ c_s k_T \rho_N p_s \sqrt{1 - \left(\frac{\frac{p}{p_s} - b}{1-b}\right)^2} & if\ b < \frac{p}{p_s} < 1 \\ -c_s k_T \rho_N p \sqrt{1 - \left(\frac{\frac{p_s}{p} - b}{1-b}\right)^2} & if\ 1 < \frac{p}{p_s} < \frac{1}{b} \\ -c_s k_T \rho_N p & if\ \frac{p}{p_s} > \frac{1}{b} \end{cases} \tag{9}$$

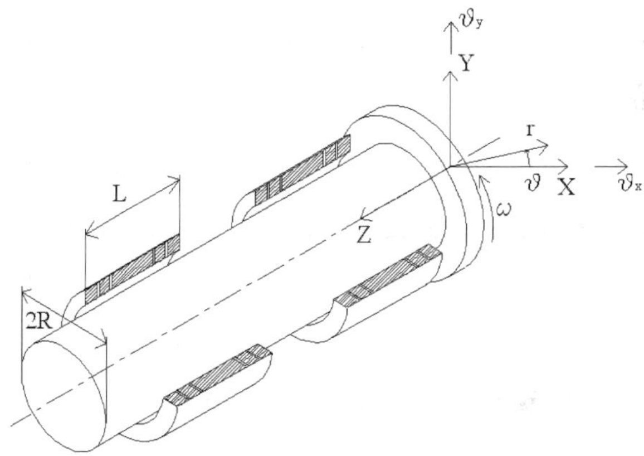

Figure 33. Schematic diagram of journal bearings

Conductance c_s appearing in the mass flow rate formula is expressed by equation (10),

$$c_s = 0.686 \frac{c_d S}{\rho_N \sqrt{R^0 T^0}} \tag{10}$$

where c_d is the discharge coefficient and S the cross-section of the supply orifice.

Discharge coefficient c_d depends on local clearance h, supply port diameter d_s and Reynolds number Re of supply port section. This relationship is taken into account with equation (11), see ref. [19].

$$c_d = 0.85\left(1 - e^{-8.2\frac{h}{d_s}}\right)\left(1 - e^{-0.005Re}\right) \tag{11}$$

The Reynolds number is calculated with eq. (12)

$$Re = \frac{4G}{\pi d_s \mu} \tag{12}$$

Equation (11) is extrapolated from experimental measurements of air consumption and pressure distributions under pads for different air gap height and supply port sizes. The equation is extrapolated in the range of h/d_s values from 12.5·10⁻³ to 100·10⁻³.

The Reynolds equation is discretized spatially by central finite-differences:

$$3h_{i,j}^2\left(\frac{\partial h}{\partial z}\right)_{i,j}\frac{p_{i+1,j}^2-p_{i-1,j}^2}{2\Delta z} + h_{i,j}^3\frac{p_{i+1,j}^2-2p_{i,j}^2+p_{i-1,j}^2}{(\Delta z)^2} + 3h_{i,j}^2\left(\frac{\partial h}{R\partial\vartheta}\right)_{i,j}\frac{p_{i,j+1}^2-p_{i,j-1}^2}{2R\Delta\vartheta} + h_{i,j}^3\frac{p_{i,j+1}^2-2p_{i,j}^2+p_{i,j-1}^2}{(R\Delta\vartheta)^2} +$$
$$24R^0T^0\mu q_{i,j} = 12\mu\omega h_{i,j}\frac{p_{i,j+1}-p_{i,j-1}}{2\Delta\vartheta} + 12\mu\omega p_{i,j}\left(\frac{\partial h}{\partial\vartheta}\right)_{i,j} + 24\mu h_{i,j}^n\frac{p_{i,j}^{n+1}-p_{i,j}^n}{\Delta t} + 24\mu p_{i,j}^n\frac{h_{i,j}^{n+1}-h_{i,j}^n}{\Delta t} \tag{13}$$

The journal equations of motion are considered together with the Reynolds equation in order to study the dynamics of the rotor-bearings system.

The 4 d.o.f. model of the rotor is described by the following system

$$\begin{cases} m_r\ddot{x}_G = F_{cx} + F_x + m_r\varepsilon\omega^2 cos(\omega t) \\ m_r\ddot{y}_G = F_{cy} + F_y + m_r\varepsilon\omega^2 sin(\omega t) \\ J_G\ddot{\vartheta}_x = M_{cx} + F_y(z_G - z_F) - J_P\omega\dot{\vartheta}_y + \chi(J_P - J_G)\omega^2 cos(\omega t - \varphi) \\ J_G\ddot{\vartheta}_y = M_{cy} - F_x(z_G - z_F) + J_P\omega\dot{\vartheta}_x + \chi(J_P - J_G)\omega^2 sin(\omega t - \varphi) \end{cases} \tag{14}$$

where J_G is the rotor transverse inertia moment, calculated with respect to the rotor centre of mass, while J_P is the polar inertia moment. The external forces have a resultant of radial components F_x, F_y applied in correspondence of $z=z_F$.

The fluid forces, comprehensive of pressure and shear actions, are expressed by

$$\begin{cases} F_{cx} = \int_0^L\int_0^{2\pi}(pcos\vartheta - \tau_\vartheta sin\vartheta)Rd\vartheta dz \\ F_{cy} = \int_0^L\int_0^{2\pi}(psin\vartheta + \tau_\vartheta cos\vartheta)Rd\vartheta dz \\ M_{cx} = \int_0^L\int_0^{2\pi}(-psin\vartheta + \tau_\vartheta cos\vartheta)(z_G - z)Rd\vartheta dz \\ M_{cy} = \int_0^L\int_0^{2\pi}(pcos\vartheta + \tau_\vartheta sin\vartheta)(z_G - z)Rd\vartheta dz \end{cases} \tag{15}$$

The tangential actions are given by:

$$\tau_\vartheta = -\frac{h}{2}\frac{\partial p}{R\partial\vartheta} - \frac{\mu\omega R}{h} \tag{16}$$

The system is solved using Euler explicit method. From relation (17) the pressure for each node is calculated at iteration n+1.

$$p_{i,j}^{n+1} = p_{i,j}^n + \Delta t\, f\left(p_{i,j}^n, p_{i+1,j}^n, p_{i-1,j}^n, p_{i,j+1}^n, p_{i,j-1}^n, h_{i,j}^n, h_{i,j}^{n-1}, \left(\frac{\partial h}{\partial\vartheta}\right)_{i,j}^n, \left(\frac{\partial h}{\partial z}\right)_{i,j}^n\right) \tag{17}$$

The explicit method has a very rapid execution per time step but is limited to the use of small Δt, see [31]. The system of nxm equations (17) is solved together with rotor equations of motion (14).

The solution procedure starts with a set of input data (shaft diameter, radial clearance, bearing axial length, position and diameter of supply orifices, shaft speed).

To calculate the static pressure distribution, h is maintained constant in time and the system is solved with initial condition $p_{i,j}=p_a$ for each node. The rotor trajectory is determined starting with the initial static pressure distribution and using the following set of initial conditions:

$$x(0) = h_0 \varepsilon_x(0);\ y(0) = h_0 \varepsilon_y(0)$$

$$\dot{x}(0) = h_0 \dot{\varepsilon}_x(0);\ \dot{y}(0) = h_0 \dot{\varepsilon}_y(0)$$

The whirl stability can be verified with the orbit method [18,32], that considers the nonlinearities of the problem.

To analyze the dynamic properties of the thrust bearing the Reynolds equation (8) is solved in time together with the axial spindle equation of motion (18):

$$m_r \ddot{z} + F_t = F \tag{18}$$

in which F_t is the reaction force of the thrust bearing, F is the external axial force applied to the shaft and m_r is the rotor mass. The reaction force is calculated by integrating the pressure over the thrust surface.

7. Conclusions

This paper investigated four prototypes of high speed spindles and rotors supported by gas bearings. The radial and axial bearings were designed using a numerical program that simulates the pressure distribution inside the air clearance both in static and in dynamic conditions.

Different were the design priorities: stiffness and load capacity for the pneumatic spindle and the electro-spindle and stability for the textile spindle and the mesoscopic spindle. Two different methods were employed to increase stability: the introduction of external damping in the textile spindle and the modification of the film geometry in the mesoscopic spindle.

The experimental activity carried out demonstrated the suitability of gas bearings for different applications and made it possible to verify the numerical models. Experimental testing played an essential role in identifying the numerical models and in the same time the models, once identified, made it possible to save time in the design process. In particular the supply holes configuration of the prototypes was optimized using the numerical models to choose their number, diameter and disposition. For stability optimization of the mesoscopic spindle the models were used as virtual test benches to compare different geometries of bushing with non-circular profile. The results of the tests on rubber rings were used in numerical models to investigate their effect on stability.

The prototypes developed operated in stable conditions in the speed range expected. Future investigations will verify the stability at higher speeds.

8. Nomenclature

b	ratio of critical pressure to admission pressure, b=0.528
c_d	supply hole discharge coefficient
c_{OR}	damping coefficient of the O-ring
c_s	supply hole conductance
D	journal bearing diameter
d_s	supply hole diameter
F_t	external axial force on thrust bearing
F_x ,F_y	external forces on rotor
G	air mass flow rate through the supply hole
h	local air clearance
h_0	clearance with rotor in centred position
J_G	transverse rotor moment of inertia, calculated respect to the centre of mass
J_P	polar moment of inertia of rotor
k_T	temperature coefficient, $k_T=\sqrt{293/T^0}$
k_{OR}	stiffness coefficient of the O-ring
L	bearing axial length
m_b	bushing mass
m_r	rotor mass
n,m	number of nodes along axial and circumferential directions
p	pressure
p_a	ambient pressure
p_s	bearing supply pressure
q	inlet mass flow rate per unit surface
R	journal bearing radius
r,θ,z	cylindrical coordinates
R^0	gas constant, in calculations R^0=287.6 m^2/s^2K
Re	Reynolds number calculated at supply port section, $Re=4G/(\pi d_s \mu)$
Re^*	modified Reynolds number, $Re^*=\varrho\omega h_0^2/\mu$
S	supply hole cross section
T^0	absolute temperature, in calculations T^0=288 K
u,v	mean velocity components in z- and θ-direction
x,y,z	cartesian coordinates
z_G	center of mass axial coordinate
z_F	axial coordinate of the external force on the rotor
e_x, e_y	rotor eccentricities
Δt	time step
Δz	mesh size in axial direction
$\Delta\theta$	mesh size in circumferential direction
Λ	bearing number, $\Lambda=6\mu\omega/p_a\cdot(D/2h_0)^2$
χ	dynamic rotor unbalance

ε	static rotor unbalance
ε_x, ε_y	rotor eccentricity ratios
γ	whirl ratio, $\gamma=\upsilon/\omega$
φ	angle between static and dynamic unbalance
μ	dynamic viscosity, in calculations $\mu=17.89\cdot10^{-6}$ Pa·s
υ	whirl frequency
ρ_N	air density in normal conditions
ω	rotor angular speed

Author details

G. Belforte, F. Colombo, T. Raparelli, A. Trivella and V. Viktorov
Department of Mechanical and Aerospace Engineering, Politecnico di Torino, Italy

9. References

[1] Fuller, D.D. (1984), *Theory and practice of lubrication for engineers*, John Wiley and Sons, New York.

[2] Waumans, T.; Peirs, J.; Al-Bender, F.; Reynaerts, D. (2011), Aerodynamic bearing with a flexible, damped support operating at 7.2 million DN, *J. Micromech. Microeng.* Vol. 21, 104014.

[3] Boffey, D. A.; Desay D. M. (1980). An experimental investigation into the rubber-stabilization of an externally-pressurized air-lubricated thrust bearing, *ASME Trans. Journal of lubrication technology*, Vol. 102, pp. 65-70.

[4] Czolczynski K. (1994), Stability of flexibly mounted self acting gas journal bearings, *Nonlinear Science, Part B, Chaos and nonlinear Mechanics*, No. 7, pp. 286-299.

[5] Zhang, R.; Chang H. S. (1995). A new type of hydrostatic/hydrodynamic gas journal bearing and its optimization for maximum stability, *STLE Tribology Transactions*, Vol. 38, No. 3, pp. 589-594.

[6] Yang, D-W; Chen, C-H; Kang, Y.; Hwang, R-M.; Shyr, S-S. (2009). Influence of orifices on stability of rotor-aerostatic bearing system, *Tribology International*, Vol. 42, pp. 1206-1219.

[7] Westwind (2008a). Online available: http://www.westwind-airbearings.com/specialist/ wafer Grinding.html

[8] Westwind (2008b). Online available: http://www.westwind-airbearings.com/pcb/overview.html

[9] APT, Air Bearing Precision Technology. Catholic University of Leuven, Belgium. Online available: http://www.mech.kuleuven.be/industry/spin/APT/default_en.

[10] Ohishi, S.; Matsuzaki, Y. (2002). Experimental investigation of air spindle unit thermal characteristics. *Precision Engineering*, Vol. 26, pp. 49-57.

[11] Moore Precision Tools (2001). *Nanotechnology Systems.*

[12] Precitech Precision, (2001). *Nanoform® 350 Technical Overview and Unsurpassed Part Cutting Results.*

[13] Toshiba Machine Co. Ltd. (2002). *High Precision Aspheric Surface Grinder.*

[14] Belforte, G.; Colombo, F.; Raparelli, T.; Trivella, A; Viktorov, V. (2008a). High speed electrospindle running on air bearings: design and experimental verification, *Meccanica*, Vol. 43, pp. 591-600.

[15] Belforte, G.; Colombo, F.; Raparelli, T.; Viktorov, V.; Trivella, A. (2006). An experimental study of high speed rotor supported by air bearings: test rig and first experimental results, *Tribology International*, Vol. 39, pp. 839-845.

[16] Della Pietra, L.; Adiletta, G. (2002). The squeeze film damper over four decades of investigations: part 1. Characteristics and operating features, *Shock Vib. Dig.*, vol. 34, pp. 3-26.

[17] Belforte, G.; Colombo, F.; Raparelli, T.; Trivella, A., Viktorov, V. (2008b). High speed rotors with air bearings mounted on flexible supports: test bench and experimental results. *ASME Journal of Tribology*, Vol. 130, 1-7.

[18] Belforte, G.; Raparelli, T.; Viktorov, V. (1999). Theoretical investigation of fluid inertia effects and stability of self-acting gas journal bearings. *ASME Journal of Tribology*, Vol. 121, 836-843.

[19] Belforte, G.; Raparelli, T.; Viktorov, V.; Trivella, A. (2007). Discharge coefficients of orifice-type restrictor for aerostatic bearings, *Tribology International*, Vol. 40, pp. 512-521.

[20] Mishra, P. C.; Pandley R. K.; Athre K. (2007). Temperature profile of an elliptic bore journal bearing, *Tribology International*, Vol. 40, pp. 453-458.

[21] Hashimoto, H. (1992). Dynamic characteristic analysis of short elliptical journal bearings in turbulent inertial flow regime, *STLE Tribology Transactions*, Vol. 35, No. 4, pp. 619-626.

[22] Hashimoto, H.; Matsumoto K. (2001). Improvement of operating characteristics of high speed hydrodynamic journal bearings by optimum design: Part I – formulation of methodology and its application to elliptical bearing design, *ASME Journal of Tribology*, Vol. 123, pp. 305-312.

[23] Wang N. Z.; Ho C. L.; Cha K. C. (2000). Engineering optimum design of fluid film lubricated bearings, *Journal of Tribology Transactions*, Vol. 43, No. 3, pp. 377-386.

[24] Read, L.J.; Flack, R.D. (1987). Temperature, pressure and film thickness measurements for an offset half bearing. *Wear*; Vol. 117, No. 2, pp. 197–210.

[25] Ene, N. M; Dimofte, F. & Keith Jr., T. G. (2008a). A dynamic analysis of hydrodynamic wave journal bearings, *STLE Tribology Transactions*, Vol. 51, No. 1, pp. 82-91.

[26] Ene, N. M; Dimofte, F. & Keith Jr., T. G. (2008b). A stability analysis for a hydrodynamic three-wave journal bearing, *Tribology International*, Vol. 41, No. 5, pp. 434-442.

[27] Sehgal, R; Swamy, K.N.S.; Athre K.; Biswas S. (2000). A comparative study of the thermal behaviour of circular and non-circular journal bearings. *Lub Sci*, Vol. 12, No. 4, pp. 329–44.

[28] Dimofte, F. (1995a). Wave journal bearing with compressible lubricant – Part I : The wave bearing concept and a comparison to the plain circular bearing, *Tribology Transactions*, Vol. 38, No. 1, pp. 153-160.

[29] Dimofte, F. (1995b). Wave journal bearing with compressible lubricant – Part II : A comparison of the wave bearing with a groove bearing and a lobe bearing, *Tribology Transactions*, Vol. 38, No. 2, pp. 364-372.

[30] Viktorov, V.; Belforte, G.; Raparelli, T., Colombo, F. (2009). Design of non-circular gas bearings for ultra-high speed spindle. *World Tribology Congress*, Kyoto, 6-11 Sept. C1-212.

[31] Castelli, V.; Pirviks, J. (1968) Review of numerical methods in gas bearing film analysis. *Journal of Lubrication Technology*, pp. 777-792.

[32] Colombo, F.; Raparelli, T.; Viktorov, V. (2009). Externally pressurized gas bearings: a comparison between two supply holes configurations. *Tribology International*, Vol. 42, 303-310.

Chapter 7

High Temperature Self-Lubricating Materials

Qinling Bi, Shengyu Zhu and Weimin Liu

Additional information is available at the end of the chapter

http://dx.doi.org/10.5772/55645

1. Introduction

There is an ongoing need for developing high temperature self-lubricating materials to meet the severe conditions of mechanical systems, such as advanced engines which require increasingly high working temperatures (at 1000 °C or above) and long life [1-7]. However, achieving and maintaining low friction and wear at high temperatures have been very difficult in the past and still are the toughest problems encountered in the field of tribology [8,9]. Yet, the efforts to explore novel high temperature self-lubricating materials possessing favorable frictional property and superior wear resistance abilities have never stopped. As a result, great strides have been made in recent years in the fabrication and diverse utilization of new high temperature self-lubricating materials that are capable of satisfying the multifunctional needs of more advanced mechanical systems [10-15]. The following tribological issues addressed in this chapter are presented:

i. High temperature self-lubricating alloys
ii. Ni matrix high temperature self-lubricating composites
iii. Intermetallics matrix high temperature self-lubricating composites
iv. Ceramic matrix high temperature self-lubricating composites
v. High temperature self-lubricating coatings.

2. High temperature self-lubricating alloys

It is well known that the friction and wear of metal alloys at high temperatures are controlled by their tribochemically generated oxide films [16-20]. Consequently, on a hard metal substrate-formed lubricating soft oxide layer, with low shear strength, results in considerable wear reduction and sometimes a decrease of friction. Hence the idea is reasonable and feasible that realization of self-lubricating property by in situ oxide formation on sliding surface at high temperatures. Peterson M.B. and Li S.Z. applied this concept to develop high temperature self-lubricating alloys, such as Ni-Cu-Re, Co-Cu-Re, and Fe-Re, by lubrication with naturally occurring oxides during the sliding process [19-23].

Meanwhile, the principles of oxide lubrication and to develop alloys based on tribochemically generated oxide films were proposed as to what alloy compositions will produce effective oxide films, what interface temperatures and what operating conditions are necessary and what oxides will be effective[20].

In addition, it is another effective approach to improvement of tribological properties by addition of active elements, such as sulfur and selenium [24,25]. At the interface, heating and sliding produces certain compounds with lubricious properties by tribochemical reaction between the active elements and metal components.

3. Ni matrix high temperature self-lubricating composites

There is a great need in current technology for solid lubricated systems that will perform satisfactorily over a wide range of temperatures. Ni matrix high temperature self-lubricating composites play a significant role in attaining this goal and achieve applications. In the past years, a series of self-lubricating composites based on nickel alloys have been developed by powder metallurgy methods [26-35]. Of these materials, PM212, which was developed by NASA Glenn (previous NASA Lewis) Research Center, shows promise for use over a wide range of temperatures (ranging from room temperature to 900 °C) [26,27]. This material is comprised of metal Ni-Co binder, ceramic Cr_3C_2 matrix and solid lubricants CaF_2/BaF_2 and Ag. These three components play roles in providing binding, wear resistance and self-lubrication, respectively. Another NASA PM304 (NiCr-Cr_2O_3-Ag-BaF_2/CaF_2) possesses well tribological behavior from room temperature to 650 °C [28], however, above 800 °C, the decline in mechanical property degrades its wear resistance. Based on the design view of NASA, many efforts are made to explore new self-lubricating composites based on nickel alloys, such as Nickel alloy-graphite-Ag, Nickel alloy-WC/SiC-PbO, Nickel alloy-Ag-CeF_3, Nickel alloy-graphite-CeF_3, Nickel alloy-MoS_2-graphite, etc. [29-35]. Although the Ni-based high temperature self-lubricating composites attract much attention, their friction and wear properties over a wide temperature range are inferior to those of PM212.

4. Intermetallics matrix high temperature self-lubricating composites

Since the strong internal order and mixed (metallic and covalent/ionic) bonding, intermetallic compounds often offer a compromise between ceramic and metallic properties when hardness and/or resistance to high temperatures is important enough to sacrifice some toughness and ease of processing [36-38]. Since Aoki and Izumi reported the remarkable achievements of ductility in Ni_3Al alloys by B doping in 1979, structural intermetallics and related materials have been actively investigated. Intermetallics have given rise to various novel high temperature self-lubricating materials developments.

4.1. Ni_3Al matrix high temperature self-lubricating composites

Ni_3Al is the intermetallic compound that has been most intensively studied from both fundamental and practical points of view [39-47]. In the past years, a great deal of work has been addressed to the study of the effect of alloying elements, mechanical properties, oxidation and corrosion. The results indicated that Ni_3Al may be an excellent matrix for high temperature self-lubricating composite owing to its high temperature strength, good oxidation resistance and corrosion resistance behavior. However, till now, the tribological behavior of Ni_3Al matrix composite has not been researched systemically.

Recently, a series of Ni_3Al high temperature self-lubricating composites were developed in Lanzhou Institute of Chemical Physics, Chinese Academy of Sciences [15, 48-55]. The self-lubricating composites, which consist of Ni_3Al matrix with Cr/Mo/W, Ag and BaF_2/CaF_2 additions, exhibit the low friction coefficient and wear rate at a wide temperature range from room temperature to 1000 °C. Additionally, in order to design and fabricate high temperature self-lubricating composite with excellent tribological property from room temperature to 1000 °C and also explore the friction and wear mechanisms at high temperatures, the effects of solid lubricant and reinforcement on tribological properties of Ni_3Al matrix high temperature self-lubricating composites at a wide temperature range from room temperature to 1000 °C were investigated. The tribological behavior was studied from room temperature to 1000 °C on an HT-1000 ball-on-disk high temperature tribometer. The schematic diagram of HT-1000 ball-on-disk high-temperature tribometer is shown in Fig. 1. The rotating disk was made of the sintered sample with a size of 18.5 × 18.5 × 5 mm, and the stationary ball was the commercial Si_3N_4 or SiC ceramic ball with a diameter of 6 mm. The selected test temperatures were room temperature, 200, 400, 600, 800 and 1000 °C. The tribological tests were carried out at an applied load of 10 or 20 N, sliding speed of 0.2 m/s and testing time of 30 or 60 min. The furnace temperature, which was monitored using a thermocouple, was raised at a heating rate of 10-12 °C /min to the set point.

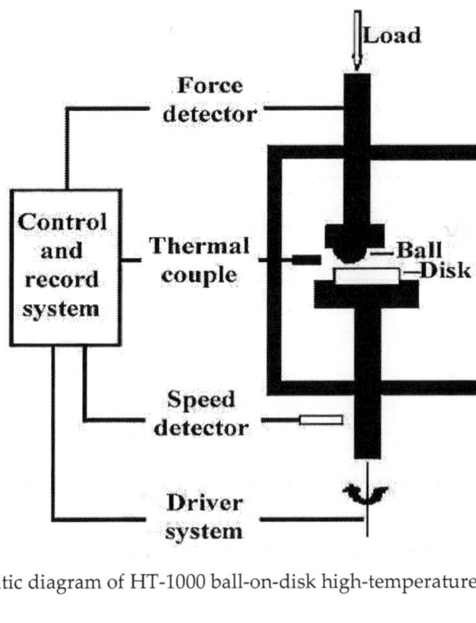

Figure 1. The schematic diagram of HT-1000 ball-on-disk high-temperature tribometer

4.1.1. *Effect of solid lubricant on the tribological behavior*

To obtain high temperature self-lubricating materials with well tribological and mechanical properties, suitable solid lubricant selected is very important. Since no single material can provide adequate lubricating properties over a wide temperature range from room temperature to high temperatures (800 or even 1000 °C), many efforts are made to a synergetic lubricating action of the composite lubricants, namely, the combination of low temperature lubricant and high temperature lubricant [56].

The conventional solid lubricants, such as MoS_2 and graphite, cannot meet the demand on tribological and mechanical properties due to their inadequate oxidation resistance in air above 500 °C . Hexagonal boron nitride (hBN) has been considered an effective solid lubricant for high temperature applications since it has a graphite-like lamellar structure. However, the non-wettability and poor sinterability of hBN would restrict its applications. Except for the above layered lubricants, soft noble metal Ag and Au should be as a promising lubricant for Ni_3Al at low temperatures (below 450 °C) due to the low shear strength and stable thermochemistry.

It was found that Ag added into the Ni_3Al matrix composite exhibited no reactants between Ag and other additives detected after the hot-sintering process. Moreover, the composite with Ag had higher strength than those with graphite or MoS_2. Furthermore, during frictional process, Ag kept favorable thermal stability at low temperatures, whereas oxidation reaction could happen between Ag and other additives in the composite at high temperatures. It is noteworthy that the oxidation products like $AgMoO_4$ are beneficial to improvement of lubricity.

In a search for even higher temperature solid lubricants for Ni_3Al, many efforts have been performed on inorganic salts and fluorides of alkali metals [51-53].

Fluorides have shown promise as high-temperature solid lubricants to provide low friction coefficient and wear according to the previous references [57,58]. Ni_3Al-Cr-Ag-BaF_2/CaF_2 composites were synthesized by powder metallurgy technique [15,51,59]. XRD results indicated that components in the sintered Ni_3Al matrix composites did not react on each other and no any new compound formed during the fabrication process. XRD patterns of worn surfaces after frictional tests presented that at 600 °C, $BaCO_3$ in the form of weak peak appears, and at 800 °C, no BaF_2 peaks present but $BaCrO_4$ peaks were found. Fluorides served as high temperature lubricants and exhibited a good reduce-friction performance at 400 and 600 °C. However, at 800 °C, $BaCrO_4$ formed on the worn surface due to the tribo-chemical reaction at high temperatures provided an excellent lubricating property.

Inorganic salts are obvious candidates for consideration owing to low shear strength and high ductility at elevated temperatures. The high temperature lubricious behavior of some sulfates, chromates, molybdates and tungstates has been extensively studied [60-66]. Important early work on high-temperature solid lubricant reported that molybdates appeared to be the promising high-temperature solid lubricants [56]. As a high-temperature solid lubricant, and similar to $CaWO_4$ and $CaMoO_4$, $BaMoO_4$ has scheelite structure and adequate thermophysial properties [67, 68]. However, till now, the lubricious behavior of $BaMoO_4$ has not been explored in detail. Recently, $BaCrO_4$ has attracted much attention due to its lubricating property at a wide temperature range [62]. $BaCrO_4$ has an orthorhombic structure, and its thermal data shows that the $BaCrO_4$ phase is thermally stable to 850 °C [69,70]. Therefore, they could be expected as promising high-temperature solid lubricants for Ni_3Al.

It can be noted that no $BaMoO_4$ peaks presented but Ni, Mo and $BaAl_2O_4$ peaks were found in XRD results of the sintered Ni_3Al matrix composites, and the peaks of Ni, Mo and $BaAl_2O_4$ get stronger with the increase of $BaMoO_4$. This means that the formation of Ni, Mo and $BaAl_2O_4$ results from high-temperature solid state reaction between Ni_3Al and $BaMoO_4$ during the fabrication process. However, during the sliding process at high temperatures, $BaMoO_4$ re-formed on the worn surfaces. The occurrence of $BaMoO_4$ is possible when considering the higher temperature rise at the instantaneous contacting surface in the rubbing process at high temperatures. It could come from the oxidation of Mo and then a reaction with $BaAl_2O_4$. The frictional results showed that Ni_3Al matrix composites with addition of $BaMoO_4$ offered better friction behavior than the monolithic Ni_3Al above 600 °C. The addition of $BaMoO_4$ could improve the tribological property, but lead to a decrease in hardness. Below 400 °C, Ni_3Al matrix composites with addition of $BaMoO_4$ wre non-lubricating, unless at 600°C, re-formed $BaMoO_4$ provided a well lubricity.

The same as $BaMoO_4$, Ni_3Al composites with addition of $BaCrO_4$ showed the absence of $BaCrO_4$ but the formation of $BaAl_2O_4$ during the fabrication process. At high temperatures, it was found the re-formation of $BaCrO_4$ on the worn surface. Since $BaMoO_4$ and $BaCrO_4$ as solid lubricants for Ni_3Al intermetallics only have low friction coefficient at narrow temperature range, they should not be used solely.

4.1.2. Effect of reinforcement on the tribological behavior

From the point of view of the principle of tribology, the ideal composition of a high temperature solid lubricant material should be composed of high strength matrix, reinforcement and solid lubricant. Reinforcement plays a significantly role in mechanical properties and tribological behavior. Generally, the reinforcement can be classified into two categories: one is the hard ceramic phase, and the other is the soft metal phase. In order to promote the tribological performance of Ni_3Al matrix composites, the different kinds of reinforcements were added.

Titanium carbide is selected as reinforcement because it is a ceramic with high melting point, extreme hardness, low density, moderate fracture toughness, and high resistance to oxidation and corrosion and a very good wettability with Ni_3Al [71-75]. Observations on TiC reinforced Ni_3Al matrix composite showed that the mechanical properties were improved, although the friction and wear performance were not promoted [59].

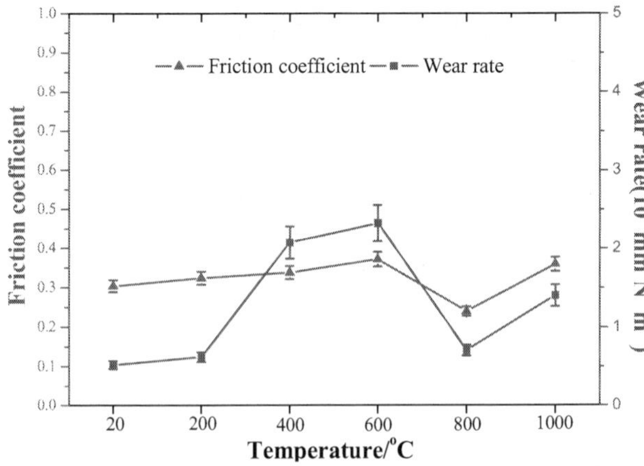

Figure 2. Variations of friction coefficients and wear rates of Ni_3Al-20%Cr-12.5%Ag-10%BaF_2/CaF_2 composite at different temperatures (tested at an applied load of 20 N and sliding speed of 0.2 m/s against Si_3N_4 ceramic ball)

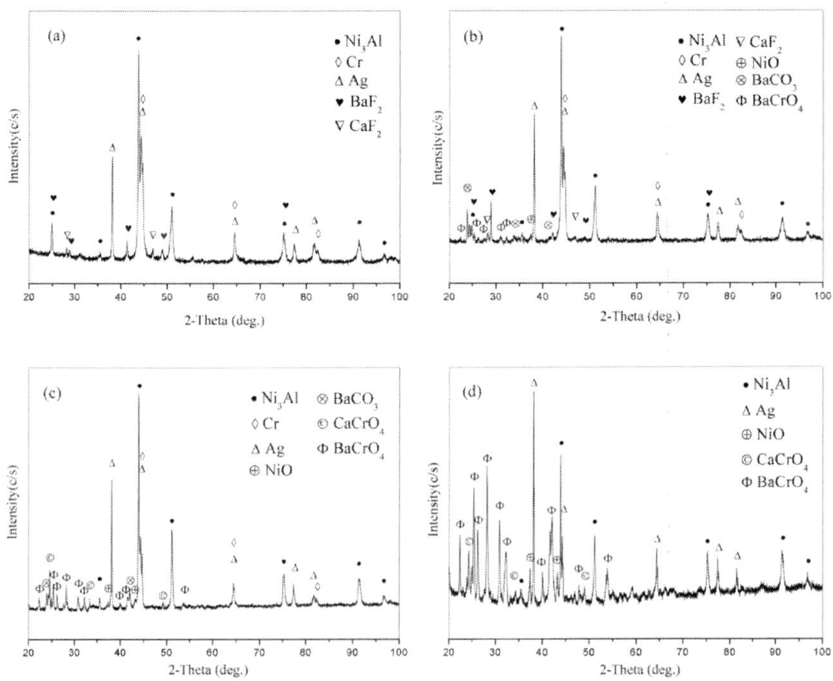

Figure 3. XRD results of the sintered sample: (a) and worn surfaces of Ni_3Al-20%Cr-12.5%Ag-10%BaF_2/CaF_2 composite after tests at different temperatures: 600 °C (b), 800 °C (c) and 1000 °C (d)

Chromium additions to Ni_3Al, as a solution, have been reported the effectiveness of alloying about 8 at% Cr for suppressing the oxygen embrittlement of Ni_3Al alloys at intermediate temperatures [39-42]. Additionally, Cr particles, as reinforcement, can improve the strength of Ni_3Al-Cr composite at low temperatures, whose strength is determined by the strength of the Cr particles and the good bonding between the matrix and Cr reinforcement [76]. The results presented that Cr added to Ni_3Al matrix composite not only enhanced mechanical strength but also ameliorated tribological performance [15]. Further study on the Ni_3Al-Cr-Ag-BaF_2/CaF_2 self-lubricating composite was carried out by tailoring the composition of the additives [48,51,59]. It was found that Ni_3Al-20%Cr-12.5%Ag-10%BaF_2/CaF_2 (in weight) composite offered the low friction coefficient 0.24-0.37 and wear rate 0.52-2.32 × 10^{-4} mm^3/Nm at a wide temperature range from room temperature to 1000 °C (shown in Fig. 2). Especially at 800 °C, the excellent self-lubricating performance was obtained among the composites.

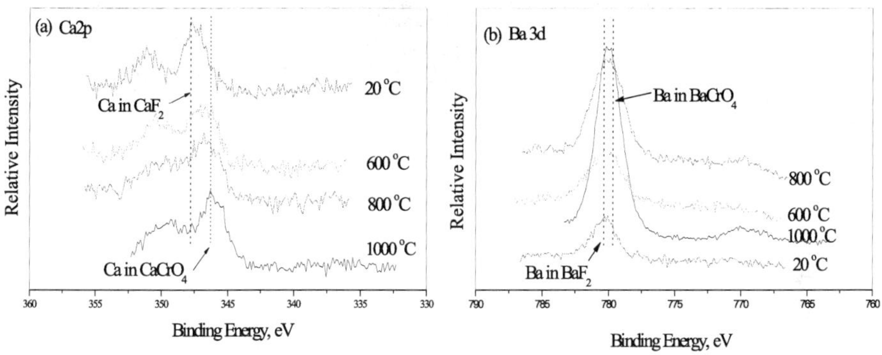

Figure 4. XPS results of worn surfaces of Ni_3Al-20%Cr-12.5%Ag-10%BaF_2/CaF_2 composite after tests at different temperatures: (a) Ca2p 3/2 photoelectron peak; (b) Ba3d 5/2 photoelectron peak

XRD results of the sintered sample and worn surfaces of Ni_3Al-20%Cr-12.5%Ag-10%BaF_2/CaF_2 composite after tests at different temperatures were represented in Fig. 3. There were no reactants among the Ni_3Al, fluorides, Ag and Cr detected after the hot-sintering process in XRD result of the sintered sample. However, peaks of $BaCO_3$ and NiO appeared on worn surface at 600 °C, and as did little $BaCrO_4$. Moreover, peaks of chromates get stronger with increase in temperature from 800 to 1000 °C, indicating that large amounts of chromates formed on worn surfaces owing to the complex reaction including high temperature reaction and tribo-chemical reaction. Also, XPS results in Fig. 4 demonstrated the formation of chromates on worn surfaces at high temperatures. The favorable self-lubricating property of Ni_3Al-BaF_2-CaF_2-Ag-Cr composite at a broad temperature range was attributed to the synergistic effects of Ag, fluorides and chromates formed at high temperatures.

Moreover, another self-lubricating composite Ni_3Al-Mo-Ag-BaF_2/CaF_2 offers acceptable mechanical strength and excellent tribological properties over a wide temperatures ranging from room temperature to 1000 °C, as shown in Table 1 and Figs. 5 and 6 [49,54,55].

Temperature/°C	20	800	900	1000
Compressive strength/MPa	1200	230	100	43

Table 1. Compressive strength of the Ni_3Al-Mo-Ag-BaF_2/CaF_2 composite at different temperatures

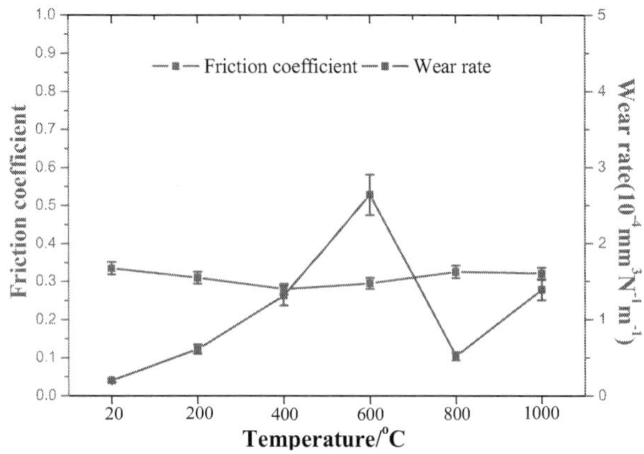

Figure 5. Variations of friction coefficients and wear rates of the Ni_3Al-Mo-Ag-BaF_2/CaF_2 composite at different temperatures (tested at an applied load of 20 N and sliding speed of 0.2 m/s against Si_3N_4 ceramic ball)

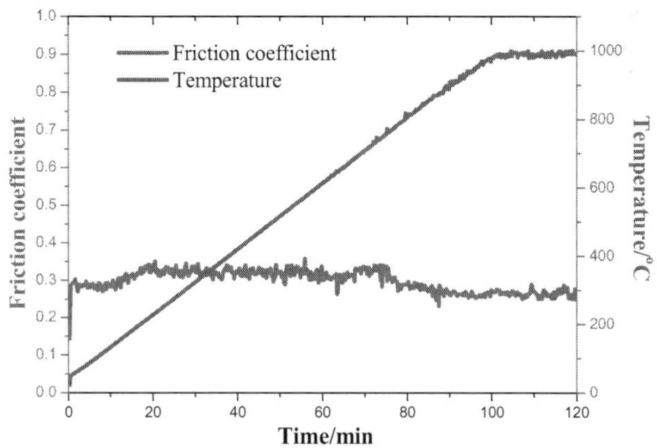

Figure 6. Evolution of friction coefficient of the Ni_3Al-Mo-Ag-BaF_2/CaF_2 composite with sliding time from room temperature to 1000 °C (tested at an applied load of 10 N and sliding speed of 0.2 m/s against Si_3N_4 ceramic ball)

In addition, tungsten as reinforcement for Ni_3Al-Ag-BaF_2/CaF_2 composite is selected based on the premise that fluoride and tungsten are expected to react with oxygen at high temperatures and create tungstate lubricants on the worn surface. As expected, barium and calcium tungstates with lubricious properties contributed to low friction coefficient at elevated temperatures [50].

4.2. NiAl matrix high temperature self-lubricating composites

Among the intermetallic family, NiAl has been selected for elevated temperature structural applications due to its low density, high oxidation resistance, high melting pointing and high conductivity [77-81]. However, NiAl is not widely used in structural applications due to its poor ductility at ambient temperatures and low strength and creep resistance at elevated temperatures. Alloying is one of effective approach that has been used successfully to improve the room temperature fracture toughness, yield strength and ductility of brittle intermetallics. NiAl-28Cr-6Mo eutectic alloys are regarded as the most logical choice of the multielement system examined to date because of their relatively high melting point, good thermal conductivity and high elevated temperature creep resistance as well as higher fracture toughness [79, 80]. Thus NiAl-28Cr-6Mo alloy may be an excellent matrix for high temperature self-lubricating composite. Recently, NiAl matrix high temperature self-lubricating composites also have been explored [82, 83]. NiAl matrix composite with various high temperature solid lubricants, such as oxide and fluoride, provide excellent lubricating properties at elevated temperatures.

It is well known that the addition of soft oxide is one of effective approach to reduce friction and wear at high temperatures because the softening oxide could offer low shear strength and high ductility and the formation of a glaze film would protect the sliding surface from heavy wear. NiAl, NiAl-Cr-Mo alloy and NiAl matrix composites with addition of oxides (ZnO/CuO) were fabricated by powder metallurgy route [82]. It was found that some new phases (such as $NiZn_3$, $Cu_{0.81}Ni_{0.19}$ and Al_2O_3) formed during the fabrication process due to a high-temperature solid state reaction. The results indicated that the monolithic NiAl had high friction coefficient and wear rate at elevated temperatures due to poor mechanical properties. The incorporation of Cr(Mo) not only enhanced mechanical properties evidently but also improved high temperature tribological properties greatly. NiAl matrix composite with addition of ZnO showed superior wear resistance at 1000 °C among the sintered materials, which was due to the formation of the ZnO layer on the worn surface. NiAl matrix composite with addition of CuO exhibited self-lubricating performance at 800 °C, which was attributed to the presence of the glaze layer containing CuO and MoO_3. Meanwhile, it had the best tribological properties among the sintered materials at 800 °C.

In addition, CaF_2 added into NiAl matrix composite exhibited favorable friction coefficient about 0.2 and excellent wear resistance about 1×10^{-5} mm^3/Nm at high temperatures (800 and 1000 °C) [83]. The excellent self-lubricating performance was attributed to the formation of the glaze film on the worn surface, which was mainly composed of $CaCrO_4$ and $CaMoO_4$ as high temperature solid lubricants. However, the composite had poor tribological performance at low temperatures. Addition of Ag evidently reduced friction coefficient and enhanced wear resistance at low temperatures. It indicated that Ag functioned as a favorable solid lubricant for NiAl intermetallic at low temperatures. However, it was adverse to friction and wear at elevated temperatures because of the decrease in the strength of material. On the whole, NiAl-Cr-Mo-CaF_2-Ag composite provided self-lubricating properties at a broad temperature range between room temperature and 1000 °C (shown in

Fig. 7) [59]. Especially at 800 °C, the composite offered excellent friction reduction about 0.2 and wear resistance about 7×10^{-5} mm^3/Nm at high temperatures. The low friction coefficient at a wide temperature range could be attributed to the synergistic effect of Ag, CaF_2, $CaCrO_4$ and $CaMoO_4$.

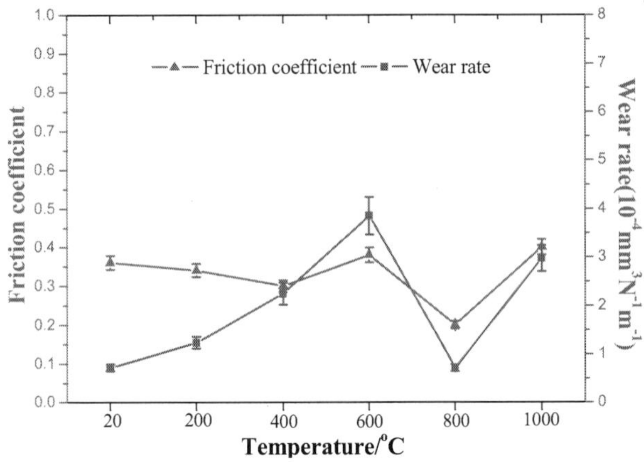

Figure 7. Variations of friction coefficients and wear rates of NiAl-Cr-Mo-CaF_2-Ag at different temperatures (tested at an applied load of 10 N and sliding speed of 0.2 m/s against Si_3N_4 ceramic ball)

5. Ceramic matrix high temperature self-lubricating composites

Advanced structural ceramics are expected to be suitable for tribo-systems because of their high hardness and corrosion resistance at high temperature [84,85]. A major challenge in advanced structural ceramics is to develop long-lifetime and reproducible ceramic sliding components for use in mechanical systems that involve high loads, velocities and temperatures. As the friction of unlubricated ceramic surfaces at elevated temperatures is usually high and unacceptable, it is necessary to find ways of effectively lubricating ceramics. Ceramic matrix composites in which solid lubricant is dispersed throughout the structure are advantageous when long lubrication life is required, compared to self-lubricating coatings. In recent years, ceramic matrix high temperature self-lubricating composites have attracted the attention of many researchers.

5.1. Zirconia matrix high temperature self-lubricating composites

Tetragonal zirconia polycrystals stabilized by yttria present a good combination of fracture toughness and bending strength, which is related to the stress-induced phase transformation of tetragonal ZrO_2 (Y_2O_3) into monoclinic symmetry. Therefore, zirconia ceramics are potential candidates for a host of engineering applications, especially at high temperatures. However, the friction coefficient of zirconia ceramics in dry sliding is enough high not to acceptable for engineering applications. Consequently, it is quite necessary to research and develop ZrO_2 (Y_2O_3) matrix high temperature self-lubricating composites.

It was reported that the additives of graphite, MoS_2, BaF_2, CaF_2, Ag, Ag_2O, Cu_2O, $BaCrO_4$, $BaSO_4$, $SrSO_4$ and $CaSiO_3$ were incorporated into zirconia ceramics, respectively, to evaluate their potentials as effective solid lubricants over a wide operating temperature range [13,86-88]. It was found that the ZrO_2 (Y_2O_3) composites incorporated with $SrSO_4$ exhibited low steady-state friction coefficients of less than 0.2 and small wear rates in the order of 10^{-6} mm^3/Nm at low sliding speed from room temperature to 800 °C. The formation, plastic deformation and effective spreading of $SrSO_4$ lubricating film were the most important factor to reduce friction and wear rate over a wide temperature range.

Recently, a ZrO_2 matrix high temperature self-lubricating composite with addition of MoS_2 and CaF_2 as lubricants prepared using hot pressing method was investigated from room temperature to 1000 °C [14,89]. The ZrO_2-MoS_2-CaF_2 composites had favorable microhardness (HV 824±90) and fracture toughness (6.5±1.4 MPa $m^{1/2}$), and against SiC ceramic exhibited excellent self-lubricating and anti-wear properties at a wide temperature range. At 1000 °C, the ZrO_2 matrix composite had a very low coefficient of friction of about 0.27 and wear rate of 1.54×10^{-5} mm^3/Nm, as shown in Figs. 8 and 9. The low friction and wear were attributed to a new lubricant $CaMoO_4$ which formed on the worn surfaces at high temperatures (seen in Fig. 10).

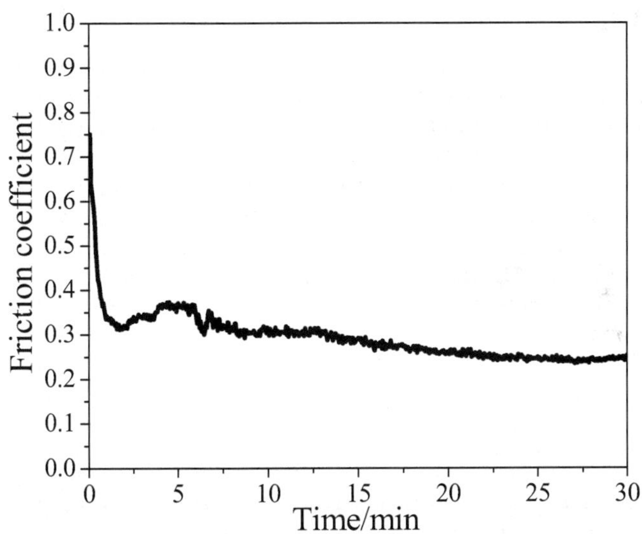

Figure 8. Evolution of friction coefficient of the ZrO_2-MoS_2-CaF_2 composite with sliding time at 1000 °C (tested at an applied load of 10 N and sliding speed of 0.2m/s against SiC ceramic ball)

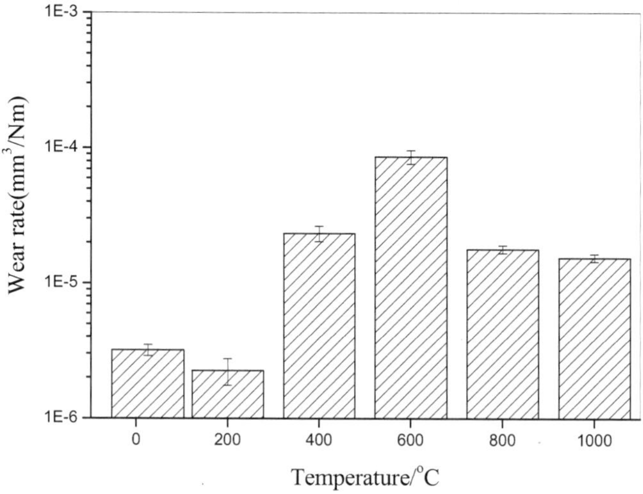

Figure 9. Variations of wear rates of the ZrO_2-MoS_2-CaF_2 composite at different temperatures (tested at an applied load of 10 N and sliding speed of 0.2m/s against SiC ceramic ball)

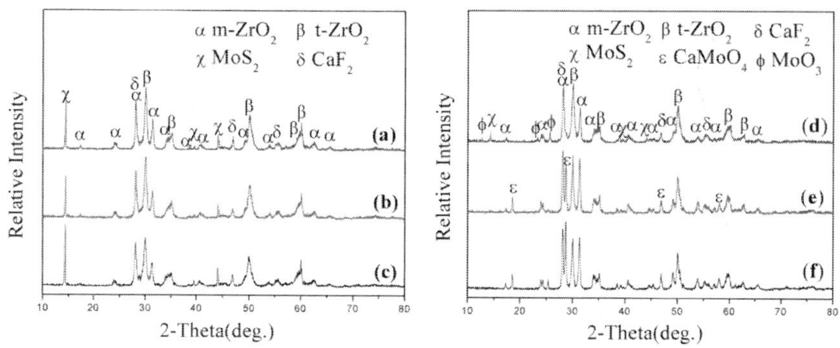

Figure 10. XRD patterns of the ZrO_2-MoS_2-CaF_2 composite (a) and its worn surfaces at different temperatures: 200 °C (b), 400 °C (c), 600 °C (d), 800 °C (e) and 1000 °C (f)

5.2. Alumina matrix high temperature self-lubricating composites

Alumina is a promising material at high temperature because of its excellent chemical stability and low price. However, tribological experiments of alumina sliding against itself at high temperature show high friction coefficient and wear rate. Solid lubrication becomes necessary to overcome this problem. In order to lubricate alumina ceramics, many efforts have been made in recent years. Among them, alumina matrix composite employed Ag and fluoride as solid lubricants is a successful example [90-92]. The Al_2O_3-Ag-CaF_2 composite exhibited a distinct improvement in wear resistance and frictional characteristics at elevated temperatures. The self-lubricating behavior was dominated by a synergistic effect. The lubricating film as a mixture of Ag and CaF_2 on friction surfaces was responsible for the reduction of friction and wear at elevated temperature.

5.3. Silicon nitride matrix high temperature self-lubricating composites

Si_3N_4-based ceramics are potential substitutes for more traditional materials for these specific applications due to their high hardness, excellent chemical and mechanical stability under a broad range of temperatures, low density, low thermal expansion and high specific stiffness [93]. The incorporation of solid lubricants is a goal to further enhance the tribological performance of Si_3N_4 [94-98]. The published papers indicated that Cs-compounds are exceptional promises as high temperature lubricants for Si_3N_4 ceramic. Cs-compound provided favorable lubrication on Si_3N_4 from room temperature to 750 °C, especially with an average value of 0.03 at 600 °C. The synergistic chemical reactions occurred between the cesium compounds, Na_2SiO_3, and the Si_3N_4 surface to provide the remarkable performance.

5.4. $M_{n+1}AX_n$ matrix high temperature self-lubricating composites

The class of refractory oxygen-free compounds possesses a layered structure and a unique combination of metal and ceramic properties, which are generally described by the formula $M_{n+1}AX_n$, where M is the transition metal, A is the preferentially subgroup IIIA or IVA element of the periodic table, and X is carbon or nitrogen [99-102]. They are characterized by a low density; high thermal conductivity, electrical conductivity, and strength; excellent corrosion resistance in aggressive external media; resistance to high-temperature oxidation; and tolerance to thermal shocks. Additionally, due to their layered structure and by analogy with hexagonal boron nitride and graphite, it is proposed that they are self-lubricating and possesses low friction coefficient. However, in the previous literature on friction and wear, $M_{n+1}AX_n$ phases did not exhibit the expected tribological properties at high temperatures. Although there exists the debate on their intrinsically self-lubricating behavior, they could be appropriate candidate for high temperature self-lubricating matrix due to the combination of metals and ceramics properties [103-107]. In order to lubricate $M_{n+1}AX_n$ phases, many efforts have been made in recent years. Among them, $M_{n+1}AX_n$ matrix composites employed Ag as solid lubricant are the promising materials for high temperature tribological applications [107-111].

6. High temperature self-lubricating coatings

High temperature self-lubricating composites with good high temperature anti-oxidation ability have been developed to reduce friction and wear from room temperature to high operating temperatures in many tribological systems. Since it is difficult or impossible for a bulk monolithic material to possess all the above mentioned surface properties [112], much attention has been paid to metallic matrix composite coatings which contain solid lubricants prepared by various processes, such as PS coatings by plasma spray [11,113], Ni/hBN composite coating by laser cladding [114], adaptive nitride-based coating by unbalanced magnetron sputtering [115], and Ni_3Al matrix composite coating by powder metallurgy [116].

6.1. PS high temperature self-lubricating coatings

In the past 40 years, the PS100, PS200, PS300 and PS400 families of plasma sprayed coatings with self-lubricating behavior were developed at NASA Lewis Research Center (shown in Table 2) [11,113,117-121]. The PS100 family of nickel-glass-solid lubricant-containing coatings pioneered the concept of combining the functions of individual constituents to produce a composite solid lubricant coating. PS200 coatings developed the composite concept, which consisted of a hard nickel-cobalt-bonded chrome carbide matrix and solid lubricants of Ag and BaF_2/CaF_2 eutectic. The PS300 coating system replaced the harder chrome carbide of PS200 coatings with chrome oxide, eliminating the necessity of costly diamond grinding and providing improved resistance to oxidative changes in high-temperature air. This coating was not very hard but had desirable tribological performance, such as good wear resistance and low friction coefficient, especially at elevated temperature up to 650 °C. NASA has recently developed a new solid lubricant coating, PS400, due to several drawbacks of PS300, namely the need to undergo a heat treatment for dimensional stabilization and poor initial surface finish. These four distinct families of coatings were engineered over the last four decades to address specific tribological challenges encountered in various aerospace systems.

Coating designation	Binder matrix	Harder	Solid lubricants	General attributes
PS100	NiCr	Glass	Ag+Fluorides	Soft-high wear
PS200	Ni-Co	Cr_3C_2	Ag+Fluorides	Hard-low wear, (abrasive to counter face dimensionally stable)
PS300	NiCr	Cr_2O_3	Ag+Fluorides	Moderate hardness, mildly abrasive to counter face, poor dimensional stability-requires heat treatment
PS400	NiMoAl	Cr_2O_3	Ag+Fluorides	Excellent dimensional stability and surface finish, poor initial low temperature tribology

Table 2. Comparision of the NASA plasma spray coating

6.2. Ni/hBN high temperature self-lubricating composite coating

The non-wettability and poor sinter ability of hBN restrict its applications as a solid lubricant, though it has a graphite-like lamellar structure. Fortunately, the hBN powders electroplated with Ni can improve the wettability with 1Cr18Ni9Ti stainless steel and sinterability as well [114]. The resulting Ni-coated hBN particulates are then used to prepare a self-lubricating wear-resistant composite coating on the stainless steel substrate with the assistance of laser cladding. Laser cladding Ni/hBN composite coating on the stainless steel substrate was composed of metallic Ni and hBN, and a small amount of B-matrix interphases, and it had high hardness and uniformly distributed constituent phases. The friction and wear behavior of the laser cladding Ni/hBN coating was strongly dependent on test temperature. The coating had good friction-reducing and anti-wear abilities as it slid against the ceramic counterpart at elevated temperatures up to 800 °C, which could be owing to the good lubricating performance of the hBN particles as a kind of high-temperature solid lubricant. The wear rate of the coating increased to some extent as the test temperature rose from 600 °C up to 800 °C, which could be attributed to the decrease in the strength of the coating at excessively high-temperature.

6.3. Adaptive nitride-based high temperature self-lubricating coatings

Adaptive tribological coatings have been recently developed as a new class of smart materials that are designed to adjust their surface chemical composition and structure as a function of changes in the working environment to minimize friction coefficient and wear between contact surfaces [12, 122-124]. At a wide temperature range, VN/Ag adaptive tribological coatings produced using unbalanced magnetron sputtering exhibited excellent self-lubricating properties [115]. The friction coefficient was found to vary from 0.35 at room temperature to about 0.15-0.20 in the 700-1000 °C range. After tribotesting, Raman spectroscopy and X-ray diffraction measurements revealed the formation of silver vanadate compounds on the surface of these coatings. In addition, real time Raman spectroscopy and high temperature XRD revealed that silver vanadate, vanadium oxide and elemental silver formed on the surface of these coatings upon heating to 1000 °C. Upon cooling, silver and vanadium oxide were found. Silver reduced the friction coefficient at low temperatures, while the Ag_3VO_4 phase provided low friction at high temperatures due to its layered atomic structure.

6.4. Intermetallics matrix high temperature self-lubricating composite coatings

Powder metallurgy is a convenient method to prepare bulk components with fine and dense microstructures. In contrast to the plasma spraying technique, where some components may be lost during the deposition process, the final composition is the same as that of the starting powders. Hence powder metallurgy is applied to prepare the coatings with a fine and dense microstructure.

One of the most attractive engineering properties of Ni_3Al alloys is their increasing yield strength with increasing temperature up to about 650-750 °C. This type of strength behavior suggests that the Ni_3Al-based intermetallic alloys may have good wear properties in the peak-strength temperature range. Consequently, investigations of Ni_3Al intermetallics for tribological coating matrix at high temperatures were initiated.

Recently, a Ni_3Al matrix coating containing Ag, Mo and BaF_2/CaF_2 was fabricated by the vacuum hot-pressed sintered technology [116]. Fig. 11 presented the morphology of the interface of the composite coating. A small amount of nickel powders were spread on the surface of the substrate to improve the wettability between the coating and AISI 1045 carbon steel. The Ni_3Al layer on the layer of nickel powder reduced the stress concentration between the substrate and the coating as well as to improve the bonding strength. The mixed composite powders Ni_3Al-Mo-Ag-fluorides were spread on the layer. After sintering, it could be seen that there were no pores and cracks near the interface region and the composite coating layer was well adhered to the substrate (Fig. 11a). The morphology of the interface of the coating (after thermal shock) was shown in Fig. 11b. It showed that the coating did not peel off and even no cracks after the thermal shock, indicating that the coating possesses excellent bonding strength.

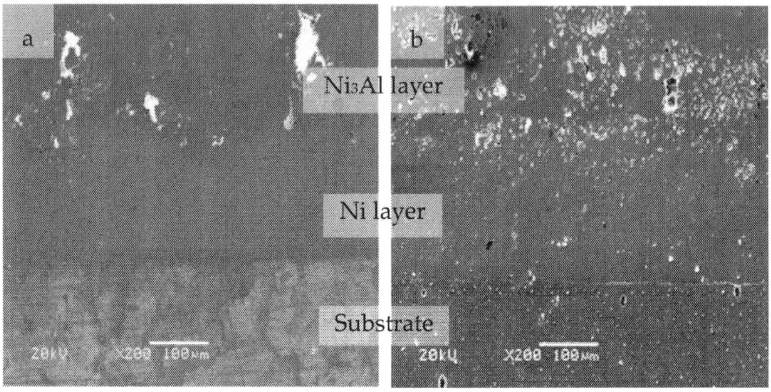

Figure 11. SEM micrographs of transverse cross-sections of the Ni_3Al-based composite coating: (a) before thermal shock; (b) after thermal shock.

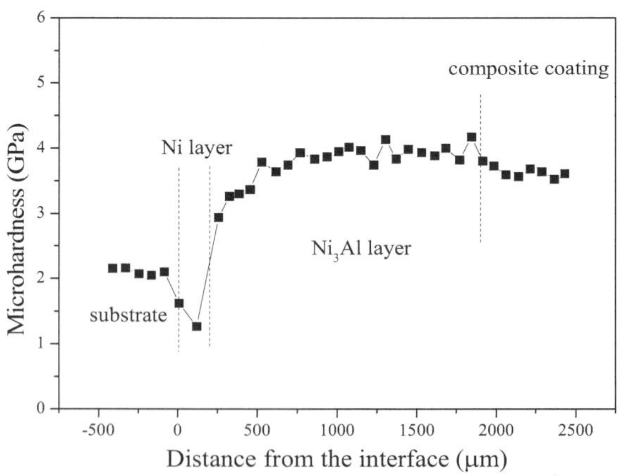

Figure 12. Microhardness profile of the Ni_3Al-based composite coating

The microhardness depth profile in the transverse cross-section of the composite coating was shown in Fig. 12. It was found that the coating was in three layers according to hardness. The Vickers hardness (HV) of the nickel interlayer was 1.40±0.50 GPa, and that of the Ni_3Al layer is 3.80±0.50 GPa, and at the top surface, the HV is 3.60±0.50 GPa. The result showed that the top layer was not the hardest part of the coating, and the hard Ni_3Al layer provided efficient support in the coating. The multilayer structure of the coating could reduce the stress concentration, and improve the bonding strength.

The average friction coefficients of the coating were presented in Fig. 13. The friction coefficients of the coating were approximately 0.35 from 25 to 800 °C, however, when the temperature reached 1000 °C, the friction coefficients fell to 0.24. In comparison, the friction coefficient of the AISI 321 stainless steel was much higher than that of the coating. Fig. 14 showed the wear rates of the coating at various temperatures in air. Although the wear rates of the coating were higher than that of the AISI 321 stainless steel at the temperature between 200 and 400 °C, but they were lower at room temperature and high temperature (above 600 °C) and remained a stable level.

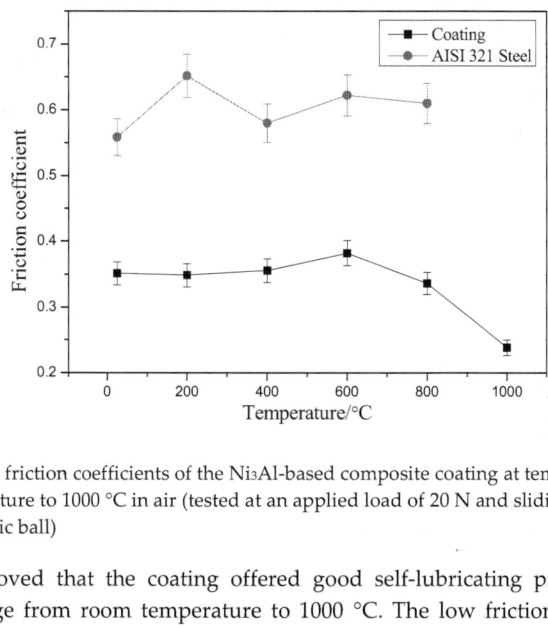

Figure 13. Average friction coefficients of the Ni_3Al-based composite coating at temperatures ranging from room temperature to 1000 °C in air (tested at an applied load of 20 N and sliding speed of 0.2 m/s against Si_3N_4 ceramic ball)

These results proved that the coating offered good self-lubricating property at a wide temperature range from room temperature to 1000 °C. The low friction coefficient of the coating was mainly attributed to Ag and fluorides eutectic at the temperature below 800 °C; at high temperatures, the molybdates, which formed in the tribochemical reaction, acted as effective lubricants (seen in Figs. 15 and 16).

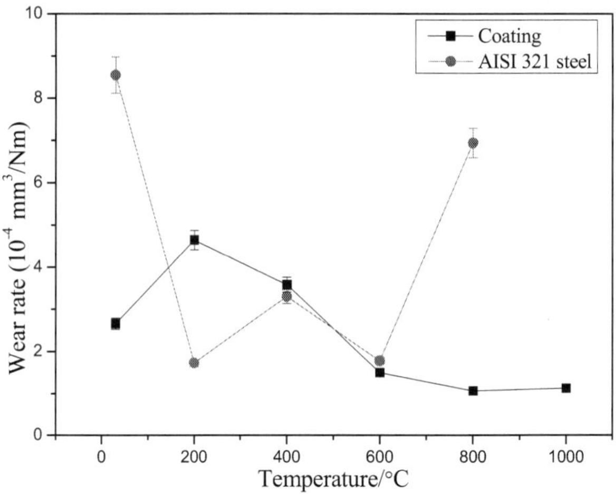

Figure 14. Wear rates of the Ni_3Al-based composite coating at temperatures ranging from room temperature to 1000 °C in air (tested at an applied load of 20 N and sliding speed of 0.2 m/s against Si_3N_4 ceramic ball)

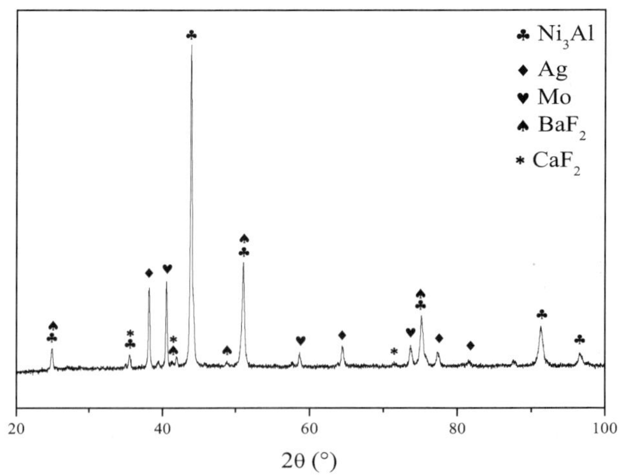

Figure 15. XRD patterns of the sintered Ni_3Al-based self-lubricating composite coating

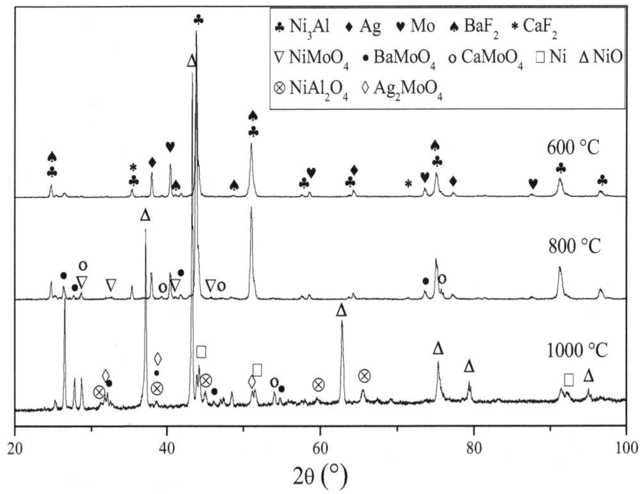

Figure 16. XRD pattern worn surfaces of the Ni3Al-based composite coating at different temperatures

7. Conclusions

There is no doubt that search for newer and better high temperature self-lubricating materials will continue in coming years, since the application conditions of future mechanical systems will undoubtedly be much more demanding than the current ones.

To fulfill engineering application, a general design appraisal of high temperature self-lubricating material can be proposed as follows: low friction (friction coefficient < 0.2); high wear resistance (wear rate < 10^{-6} mm^3/Nm); and wide temperature range (from room temperature to high temperature of above 1000 °C).

To meet these requirements, matrix and solid lubricant selected are essential for novel high temperature self-lubricating material.

Intermetallic alloys own the comprehensive mechanical properties for industrial applications by micro- or macro-alloying process. They are now serious candidates for structural applications requiring reduced density, excellent fatigue strength, corrosion/oxidation resistance, and service at temperatures up to 1000 °C. Intermetallic alloys have attracted the attention of many researchers as promising tribomaterials for mechanical components operating under severe or hostile environments.

It should be realized by designers and engineers that there is no "universal lubricant" that can operate at a broad temperature range conditions. A synergetic lubricating action, a mixture of two or more solid lubricants, is one of promising approaches to fabricate high temperature self-lubricating materials. Furthermore, research on novel solid lubricant is also a response to the requirements.

Author details

Qinling Bi, Shengyu Zhu and Weimin Liu
State Key Laboratory of Solid Lubrication, Lanzhou Institute of Chemical Physics, Chinese Academy of Sciences, Lanzhou, PR China

Acknowledgement

The authors are grateful to the National Natural Science Foundation of China (51075383) and the National 973 Project of China (2013CB632300) for financial support. And also, we thank Dr. Muye Niu and Lingqian Kong for their help to the paper.

8. References

[1] Liu WM, Xue QJ. New direction in tribology. China mechanical engineering 2000, 11: 77-80.

[2] Xue QJ, Lu JJ. Research Status and Developing Trend of Solid Lubrication at High Temperatures. Tribology (in Chinese) 1998; 19: 91-6.

[3] Perry SS, Tysoe WT. Frontiers of fundamental tribological research. Tribology Letters 2005; 19: 151-61.

[4] Spikes H. Tribology research in the twenty-first century. Tribology International 2001; 34: 789-99.

[5] Hiroshi H. High Temperature Materials for Gas Turbines: The Present and Future. Proceedings of the International Gas Turbine Congress, Tokyo, 2003.

[6] Zhao JC. Ultrahigh-temperature materials for jet engines. MRS Bulletin 2003; 28: 620-30.

[7] Wright IG, Gibbons TB. Recent developments in gas turbine materials and technology and their implications for syngas firing. International Journal of Hydrogen Energy 2007; 32: 3610-21.

[8] Erdemir A. A crystal-chemical approach to lubrication by solid oxides. Tribology Letters 2000; 8: 97-102.

[9] Peterson MB, Murray SF, Florek JJ. Consideration of lubricants for Temperatures above 1000 °F. Tribology Transactions 1959; 2: 225-34.

[10] Donnet C, Erdemir A. Solid lubricant coatings: recent developments and future trends. Tribology Letters 2004; 17: 389-397.

[11] Dellacorte C, Stanford MK, Thomas F, Edmonds BJ. The Effect of Composition on the Surface Finish of PS400: A New High Temperature Solid Lubricant Coating. NASA-TM-2010-216774, 2010.

[12] Aouadi SM, Luster B, Kohli P, Muratore C, Voevodin AA. Progress in the development of adaptive nitride-based coatings for high temperature tribological applications. Surface & Coatings Technology 2009; 204: 962-968.

[13] Ouyang JH, Li YF, Wang YM, Zhou Y, Murakami T, Sasakic S. Microstructure and tribological properties of $ZrO_2(Y_2O_3)$ matrix composites doped with different solid lubricants from room temperature to 800 °C. Wear 2009; 267: 1353-60.

[14] Kong LQ, Bi QL, Niu MY, Zhu SY, Yang J, Liu WM. Journal of the European Ceramic Society 2013; 33: 51-59.

[15] Zhu SY, Bi QL, Yang J, Liu WM, Xue QJ. Ni_3Al matrix high temperature self-lubricating composites. Tribology International 2011; 44: 445-53.

[16] Woydt M, Dogigli M, Agatonovic P. Concepts and technology development of hinge joints operated up to 1600°C in air. Tribology Transactions 1997; 40: 643-6.

[17] Woydt M, Skopp A, Dörfel I, Witke K. Wear Engineering Oxides/Antiwear Oxides. Tribology Transactions 1999; 42: 21-31.

[18] Pauschitz A, Roy M, Franek F. Mechanisms of sliding wear of metals and alloys at elevated temperatures. Tribology International 2008; 41: 584-602.

[19] Peterson MB, Li SZ, Murray SF. Wear-resisting Oxide Films for 900 °C. Journal of Material Sciences and Technology 1997; 13: 99-106.

[20] Peterson MB, Caiabrese SJ, Li SZ, Jiang XX. Friction of Alloys at High Temperature. J.Mater.Sci .Technol. 1994; 10: 313-320.

[21] Li SZ, Jiang XX, Yin FC, Peterson MB, Calabrese SJ. On self-lubricating behavior of Ni - Cu-Re alloys at elevated temperature. Chinese Journal of Materials Research1989; 3: 481-486.

[22] Jiang XX, Li SZ, Peterson MB, Calabrese SJ. On self-lubricating behavior of Co-Cu-Re alloys at elevated temperature. Chinese Journal of Materials Research 1989; 3:487-93.

[23] Xiong Dangsheng;Li Xibin;Li Shizhuo;Jiang Xiaoxia Self-lubricating behavior and its relation to matching pair of Fe-Re alloy at elevated temperature. The Chinese Journal of Nonferrous Metals 1995;5:115-118

[24] Kan CY, Liu JZ, Zhang GW, Ouyang JL.Preparation and Tribological Performance o f a Ni-Cr-S Alloy. Tribology (in Chinese) 1994; 14: 193-204.

[25] Wang Y, Wang JB, Wang JA, Zhao JZ, Ouyang JL. Study on the Nickel Alloy Containing Sulfide and Its Tribological Properties at High Temperature. Tribology (in Chinese) 1996; 16: 289-97.

[26] Dellacorte C, Sliney HE. Tribological properties of PM212-a high-temperature, self-lubricating, powder-metallurgy composite. Lubrication Engineering 1991; 47: 298-303.

[27] Dellacorte C, Sliney HE. Tribological and Mechanical Comparison of sintered and HIPped PM212: high temperature self-lubricating composites. Lubrication Engineering 1992; 48: 877-85.

[28] Ding CH, Li PL, Ran G, Tian YW, Zhou JN. Tribological property of self-lubricating PM304 composite. Wear 2007; 262: 575-81.

[29] Lu JJ, Xue QJ, Wang JB. The effect of CeF_3 on the mechanical and tribological properties of Ni-based alloy. Tribology International 1997; 30: 659-62.

[30] Lu JJ, Xue QJ, Zhang GW. Effect of silver on the sliding friction and wear behavior of CeF_3 compact at evaluated temperatures. Wear 1998; 214: 103-11.

[31] Niu SQ, Zhu JP, Ouyang JL. Study on the development and properties of several new type high temperature self-lubricating composites. Tribology (in Chinese) 1995; 15: 324-32.

[32] Lu JJ, Yang SR, Wang JB, Xue QJ. Mechanical and tribological properties of Ni-based alloy/CeF_3/graphite high temperature self-lubricating composites. Wear 2001; 249: 1070-6.

[33] Li JL, Xiong DS. Tribological properties of nickel-based self-lubricating composite at elevated temperature and counterface material selection. Wear 2008; 265: 533-9.

[34] Wang JB, Lu JJ, Ouyang JL, Xue QJ. Study on the Tribological Properties of SiC-Ni-Co-Mo-PbO High-Temperature Self-Lubricating Cermet Material. Tribology (in Chinese) 1994; 14: 49-56.

[35] Wang JB, Huang YZ, Ouyang JL. Study on the tribological of WC-Ni-Co-Mo-PbO high-temperature self-lubricating cermet material. Tribology (in Chinese) 1995; 3: 205-10.Zhang YG, Han YF, Chen GL, Guo JT, Wang XJ, Feng D. Structural Intermetallics. Beijing, National Defence Industry Press, 2001.

[37] Deevi SC, Sikkat VK, Liu CT. Processing, properties, and applications of nickel and iron aluminides. Progress in Materials Science 1991; 42: 177-92.

[38] Yamaguchi M, Inui H, Ito K. High temperature structural intermetallics. Acta Materialia 2000; 48: 307-22.

[39] Liu CT, White CL. Dynamic embrittlement of boron-doped Ni_3Al alloys at 600°C. Acta Metallurgica 1987; 35: 643-9.

[40] Czeppe T, Wierzbinski S. Structure and mechanical properties of NiAl and Ni_3Al-based alloys. International Journal of Mechanical Sciences 2000; 42: 1499-518.

[41] Sikka VK, Deevi SC, Viswanathan S, Swindeman RW, Santella ML. Advances in processing of Ni_3Al-based intermetallics and applications. Intermetallics 2000; 8: 1329-37.

[42] Lapin J. Effect of ageing on the microstructure and mechanical behaviour of a directionally solidified Ni_3Al-based alloy. Intermetallics 1997; 5: 615-24.

[43] Lee WH. Oxidation and sulfidation of Ni_3Al. Materials Chemistry and Physics 2002; 76: 26-37.

[44] Lee DB. Santella ML. High temperature oxidation of Ni_3Al alloy containing Cr, Zr, Mo, and B. Materials Science and Engineering A 2004; 374: 217-23.

[45] Pérez P, González-Carrasco JL, Adeva P. The Effect of Cr Implantation on the Isothermal-Oxidation Behavior of a Ni_3Al PM Alloy. Oxidation of Metals 1999; 51: 273-89.

[46] Blau PJ, DeVore CE. Sliding friction and wear behaviour of several nickel aluminide alloys under dry and lubricated conditions. Tribology International 1990; 23: 226-34.

[47] Goldenstein H, Silva YN, Yoshimura HN. Designing a new family of high temperature wear resistant alloys based on Ni_3Al IC: experimental results and thermodynamic modeling. Intermetallics 2004; 12: 963-8.

[48] Zhu SY, Bi QL, Yang J, Liu WM. Influence of Cr content on tribological properties of Ni_3Al matrix high temperature self-lubricating composites. Tribology International 2011; 44: 1182-7.

[49] Zhu SY, Bi QL, Yang J, Liu WM, Xue QJ. Effect of particle size on tribological behavior of Ni_3Al matrix high temperature self-lubricating composites. Tribology International 2011; 44: 1800-9.

[50] Zhu SY, Bi QL, Yang J, Liu WM. Ni_3Al matrix composite with lubricous tungstate at high temperatures. Tribology Letters, 2012, 45: 251-5.

[51] Zhu SY, Bi QL, Yang J, Liu WM. Effect of fluoride content on friction and wear performance of Ni_3Al matrix high temperature self-lubricating composites. Tribology Letters 2011; 43: 341-9.

[52] Zhu SY, Bi QL, Yang J, Liu WM. Tribological property of Ni_3Al matrix composites with addition of $BaMoO_4$. Tribology Letters 2011; 43: 55-63.

[53] Zhu SY, Bi QL, Kong LQ, Yang J, Liu WM. Barium chromate as a solid lubricant for nickel aluminum. Tribology Transactions, 2012, 55: 218-23.

[54] Zhu SY, Wang LF, Chen JT, Tian Y, Bi QL. Ni_3Al matrix high temperature self-lubricating composites at a wide temperature range. Aerospace materials and Technology (in Chinese), accept.

[55] Xu JL, Yan CQ, Zhu SY, Yang J, Bi QL. Tribological Properties of Ni_3Al Matrix Self-lubricating Composites. Tribology (in Chinese) 2012; 32:584-90.

[56] Sliney HE. Solid lubricant materials for high temperatures-a review. Tribology International 1982; 5: 303-15.

[57] Deadmore DL, Sliney HE. Hardness of CaF_2 and BaF_2 solid lubricants at 25-670 °C. NASA-TM-88979, 1987.

[58] Sliney HE, Strom TN, Allen GP. Fluoride solid lubricants for extreme temperatures and corrosive environments. NASA-TM-X-52077, 1965.

[59] Zhu SY. Fabrication and Tribological Performance of Ni-Al Matrix High Temperature Self-lubricating Composites. PhD Thesis. Chinese Academy of Sciences; 2011.

[60] John PJ, Zabinski JS. Sulfate based coatings for use as high temperature lubricants. Tribology Letters 1999; 7: 31-7.

[61] John PJ, Prasad SV, Voevodin AA, Zabinski JS. Calcium sulfate as a high temperature solid lubricant. Wear 1998; 219: 155-61.

[62] Ouyang JH, Sasaki S, Murakami T, Umeda K. Spark-plasma-sintered $ZrO_2(Y_2O_3)$-$BaCrO_4$ self-lubricating composites for high temperature tribological applications. Ceramic International 2005; 31: 543-53.

[63] Murakami T, Ouyang JH, Sasaki S, Umeda K, Yoneyama Y. High-temperature tribological properties of spark-plasma-sintered Al_2O_3 composites containing barite-type structure sulfates. Tribology International 2007; 40: 246-53.

[64] Gulbiński W, Suszko T. Thin films of MoO_3-Ag_2O binary oxides - the high temperature lubricants. Wear 2006; 261: 867-73.

[65] Gulbiński, W., Suszko, T., Sienicki, W., Warcholiński, B.: Tribological properties of silver- and copper- doped transition metal oxide coatings. Wear 2003; 254: 129-35.

[66] Prasad SV, McDevitt NT, Zabinski JS. Tribology of tungsten disulfide-nanocrystalline zinc oxide adaptive lubricant films from ambient to 500°C. Wear 2000; 237: 186-96.

[67] Wu XY, Du J, Li HB, Zhang MF, Xi BJ, Fan H, Zhu YC, Qian YT. Aqueous mineralization process to synthesize uniform shuttle-like $BaMoO_4$ microcrystals at room temperature. Journal of Solid State Chemistry 2007; 180: 3288-95.

[68] Basiev TT, Sobol, AA, Voronko YK, Zverev PG. Spontaneous Raman spectroscopy of tungstate and molybdate crystals for Raman lasers. Optical Materials 2000; 15: 205-16.

[69] Gabr RM, El-Award AM, Girgis MM. Physico-chemical and catalytic studies on the calcinations products of $BaCrO_4$-CrO_3 mixture. Materials Chemistry and Physics 1992; 30: 253-9.

[70] Azad AM, Sudha R, Sreedharam OM. The standard Gibbs energies of formation of $ACrO_4$ (A = Ca, Sr or Ba) from EMF measurements. Thermochimica Acta 1992; 194: 129-36.

[71] Tiegs TN, Alexander KB, Plucknett KP, Menchenhofer PA, Becher PF, Waters SB. Ceramic composites with a ductile Ni_3Al binder phase. Materials Science and Engineering A 1996; 209: 243-7.

[72] Wall J, Chooa H, Tiegs TN, Liaw PK. Thermal residual stress evolution in a TiC-50 vol.% Ni_3Al cermet. Materials Science and Engineering A 2006; 421: 40-5.

[73] Zhang LM, Liu J, Yuan RZ, Hirai T. Properties of TiC-Ni_3Al composites and structural optimization of TiC-Ni_3Al functionally gradient materials. Materials Science and Engineering A 1995; 203: 272-7.

[74] Becher PF, Plucknett KP. Properties of Ni_3Al-bonded Titanium Carbide Ceramics. Journal of European Ceramic Society 1997; 18: 395-400.

[75] Plucknett KP, Becher PF, Waters SB. Flexure Strength of Melt-Infiltration-Processed Titanium Carbide/Nickel Aluminide Composites. Journal of American Ceramic Society 1998; 81: 1839-44.

[76] García Barriocanal J, Pérez P, Garcés G, Adeva P. Microstructure and mechanical properties of Ni_3Al base alloy reinforced with Cr particles produced by powder metallurgy. Intermetallics 2006; 14: 456-63.

[77] Bei H, George EP. Microstructures and mechanical properties of a directionally solidified NiAl-Mo eutectic alloy. Acta Materialia 2005; 53: 69-77.

[78] Grabk HJ, Brumm MW, Wagemann B. The oxidation of NiAl, Materials and Corrosion 1996; 47: 675-7.

[79] Johnson DR, Chen XF, Oliver BF, Noebe RD, Whittenberger JD. Processing and mechanical properties of in-situ composites from the NiAl-Cr and the NiAl-(Cr, Mo) eutectic systems. Intermetallics 1995; 3: 99-113.

[80] Huai KW, Guo JT, Gao Q, Li HT, Yang R. Microstructure and mechanical behavior of NiAl-based alloy prepared by powder metallurgical route. Intermetallics 2007; 15: 749-52.

[81] Guo JT. Ordered Intermetallic Compound NiAl Alloy. Beijing: Science Press, 2003.

[82] Zhu SY, Bi QL, Niu MY, Yang J, Liu WM. Tribological behavior of NiAl matrix composites with addition of oxides at high temperatures. Wear 2012, 274-275: 423-34.

[83] Zhu SY, Bi QL, Wu HR, Yang J, Liu WM. NiAl matrix high temperature self-lubricating composite. Tribology Letters 2011; 41: 535-40.

[84] Xue QJ, Liu HW. Tribology of Ceramics: Friction and wear of Ceramics. Tribology (in Chinese) 1995; 15: 376-84.

[85] Xue QJ, Liu HW. Tribology of Ceramics: Lubrication of Ceramics. Tribology (in Chinese) 1996; 16: 184-92.

[86] Ouyang JH, Sasaki S, Murakami T, Umeda K. Tribological properties of spark-plasma-sintered $ZrO_2(Y_2O_3)$-CaF_2-Ag composites at elevated temperatures. Wear 2005; 258: 1444-54.

[87] Ouyang JH, Sasaki S, Murakami T, Umeda K. Spark-plasma-sintered $ZrO_2(Y_2O_3)$-$BaCrO_4$ self-lubricating composites for high temperature tribological applications. Ceramic International 2005; 31: 543-53.

[88] Ouyang JH, Sasaki S, Umeda K. Microstructure and tribological properties of low-pressure plasma-sprayed ZrO_2-CaF_2-Ag_2O composite coating at elevated temperature. Wear 2001; 249: 440-51.

[89] Kong LQ, Bi QL, Zhu SY, Yang J, Liu WM. Tribological properties of ZrO_2 (Y_2O_3)-Mo-BaF_2/CaF_2 composites at high temperatures. Tribology International, 2012; 45: 43-9.

[90] Jin Y, Kato K, Umehara N. Tribological properties of self-lubricating CMC/Al_2O_3 pairs at high temperature in air, Tribology Letters 1998; 4: 243-50.

[91] Jin Y, Kato K, Umehara N. Effects of sintering aids and solid lubricants on tribological behaviours of CMC/Al_2O_3 pair at 650 °C. Tribology Letters 1999; 6: 15-21.

[92] Jin Y, Kato K, Umehara N. Further investigation on the tribological behavior of Al_2O_3-20Ag-20CaF_2 composite at 650 °C. Tribology Letters 1999; 6: 225-32.

[93] Riley FL. Silicon nitride and related materials. Journal of the American Ceramic Society 2000; 83: 245-65.

[94] Skopp A, Woydt M, Habig K-H. Tribological behavior of silicon nitride materials under unlubricated sliding between 22 °C and 1000 °C. Wear 1995; 181-183: 571-80.

[95] Carrapichano JM, Gomes JR, Silva RF. Tribological behaviour of Si_3N_4-BN ceramic materials for dry sliding pplications. Wear 2002; 253: 1070-6.

[96] Strong KL, Zabinski JS. Tribology of pulsed laser deposited thin films of cesium oxythiomolybdate (Cs_2MoOS_3). Thin Solid Films 2002; 406: 174-84.

[97] Strong KL, Zabinski JS. Characterization of annealed pulsed laser deposited (PLD) thin films of cesium oxythiomolybdate (Cs_2MoOS_3). Thin Solid Films 2002; 406: 164-73.

[98] Rosado L, Forster NH, Trivedi HK, King JP. Solid Lubrication of Silicon Nitride with Cesium-Based Compounds: Part I-Rolling Contact Endurance, Friction and Wear. Tribology Transactions 2000; 43: 489-97.

[99] Nowotny VH. Strukturchemie einiger Verbindungen der Übergangsmetalle mit den elementen C, Si, Ge, Sn. Progress in Solid State Chemistry 1971; 5: 27-70.

[100] Barsoum MW, El-Raghy T. Synthesis and Characterization of a Remarkable Ceramic: Ti_3SiC_2. Journal of the American Ceramic Society 1996; 79: 1953-6.

[101] Barsoum MW. The $M_{N+1}AX_N$ phases: A new class of solids: Thermodynamically stable nanolaminates. Progress in Solid State Chemistry 2000; 28: 201-81.

[102] Eklund P, Beckers M, Jansson U, Högberg H, Hultman L. The $M_{n+1}AX_n$ phases: Materials science and thin-film processing Thin Solid Films 2010; 518: 1851-78.

[103] Myhra S, Summers JWB, Kisi EH. Ti_3SiC_2-A layered ceramic exhibiting ultra-low friction. Materials Letters 1999; 39: 6-11.

[104] Zhang Y, Ding GP, Zhou YC, Cai BC. Ti_3SiC_2-a self-lubricating ceramic. Materials Letters 2002; 55: 285-9.

[105] Zhai HX, Huang ZY, Ai MX. Tribological behaviors of bulk Ti_3SiC_2 and influences of TiC impurities. Materials Science and Engineering A 2006; 435-436: 360-70.

[106] Wan DT, Hu CF, Bao YW, Zhou YC. Effect of SiC particles on the friction and wear behavior of $Ti_3Si(Al)C_2$-based composites. Wear 2007; 262: 826-32.

[107] Ren SF, Meng JH, Lu JJ, Yang SR. Tribological Behavior of Ti_3SiC_2 Sliding Against Ni-based Alloys at Elevated Temperatures. Tribology Letters 2008; 31:129-37.

[108] Gupta S, Filimonov D, Zaitsev V, Palanisamy T, Barsoum MW. Ambient and 550 °C tribological behavior of select MAX phases against Ni-based superalloys. Wear 2008; 264: 270-8.

[109] Gupta S, Filimonov D, Palanisamy T, Barsoum MW. Tribological behavior of select MAX phases against Al_2O_3 at elevated temperatures. Wear 2008; 265: 560-5.

[110] Gupta S, Filimonov D, Palanisamy T, El-Raghy T, Barsoum MW. Ta_2AlC and Cr_2AlC Ag-based composites-New solid lubricant materials for use over a wide temperature range against Ni-based superalloys and alumina. Wear 2007; 262: 1479-89.

[111] Gupta S, Filimonov D, Palanisamy T, El-Raghy T, Barsoum MW. Study of tribofilms formed during dry sliding of Ta_2AlC/Ag or Cr_2AlC/Ag composites against Ni-based superalloys and Al_2O_3. Wear 2009; 267: 1490-500.

[112] Wang HM, Yu YL, Li SQ. Microstructure and tribological properties of laser clad CaF_2/Al_2O_3 self-lubrication wear-resistant ceramic matrix composite coatings. Scripta Materialia 2002; 47: 57-61.

[113] Dellacorte C, Edmonds BJ. Preliminary Evaluation of PS300: A New Self-Lubricating High Temperature Composite Coating for Use to 800°C. NASA-TM-107056, 1996.

[114] Zhang ST, Zhou JS, Guo BG, Zhou HD, Pu YP, Chen JM. Friction and wear behavior of laser cladding Ni/hBN self-lubricating composite coating. Materials Science and Engineering A 2008; 491: 47-54.

[115] Aouadi SM, Singh DP, Stone DS, Polychronopoulou K, Nahif F, Rebholz C, Muratore C, Voevodin AA. Adaptive VN/Ag nanocomposite coatings with lubricious behavior from 25 to 1000 °C. Acta Materialia 2010; 58: 5326-31.

[116] Niu MY, Bi QL, Yang J, Liu WM. Tribological performance of a Ni_3Al matrix self-lubricating composite coating tested from 25 to 1000 °C. Surface and Coatings Technology 2012; 206: 3938-43.

[117] Sliney HE. Wide temperature spectrum self-lubricating coatings prepared by plasma spraying. Thin Solid Films 1979; 64: 217-21.

[118] Walther GC. Program for Plasma-Sprayed Self-Lubricating Coatings. NASA-CR-3163, 1979.

[119] Dellacorte C, Edmonds BJ. NASA PS400: A New High Temperature Solid Lubricant Coating for High Temperature Wear Applications. NASA-TM-2009-215678, 2009.

[120] Heshmat H, Hryniewicz P, Walton II JF, Willis JP, Jahanmir S, Dellacorte C. Low-friction wear-resistant coatings for high-temperature foil bearings. Tribology International 2005; 38: 1059-75.

[121] Wang WC. Application of a high temperature self-lubricating composite coating on steam turbine components. Surface and Coatings Technology 2004; 177-178: 12-7.

[122] Basnyat P, Luster B, Kertzman Z, Stadler S, Kohli P, Aouadi S, Xu J, Mishra SR, Eryilmaz OL, Erdemir A. Mechanical and tribological properties of CrAlN-Ag self-lubricating films. Surface and Coatings Technology 2007; 202: 1011-6.

[123] Mulligan CP, Blanchet TA, Gall D. CrN-Ag nanocomposite coatings: High-temperature tribological response. Wear 2010; 269: 125-31.

[124] Aouadi SM, Paudel Y, Simonson WJ, Ge Q, Kohli P, Muratore C, Voevodin AA. Tribological investigation of adaptive Mo_2N/MoS_2/Ag coatings with high sulfur content. Surface and Coatings Technology 2009; 203: 1304-9.

Chapter 8

Development of Eco-Friendly Biodegradable Biolubricant Based on Jatropha Oil

M. Shahabuddin, H.H. Masjuki and M.A. Kalam

Additional information is available at the end of the chapter

http://dx.doi.org/10.5772/51376

1. Introduction

Various types of lubricants are available all over the world including mineral oils, synthetic oils, re-refined oils, and vegetable oils. Most of the lubricants which are available in the market are based on mineral oil derived from petroleum oil which are not adaptable with the environment because of its toxicity and non-biodegradability [1, 2]. Unknown petroleum reserve and the increasing consumption, which made concern to use petroleum based lubricant thus, to find the alternative lubricant to meet the future demand is an important issue [3]. Therefore, vegetable oil can be played a vital role to substitute the petroleum lubricant as it possesses numerous advantage over base lubricant like renewability, environmentally friendly, biodegradability, less toxicity and so on [4-8]. It has been reported that yearly 12 million tons of lubricants waste are released to the environment [9]. However, it is very difficult to dispose it safely for the mineral oil based lubricants due its toxic and non-biodegradable nature. To reduce the dependency on petroleum fuel, legislations have been passed to use certain percentage of biofuel in many countries, such initiative also required for lubricant as well [10]. Vegetable oils are mainly triglycerides which contain three hydroxyl groups and long chain unsaturated free fatty acids attached at the hydroxyl group by ester linkages acids favors triglycerides crystallization [11, 12]. The unsaturated free fatty acid which is defined as the ratio and position of carbon-carbon double bond, one two and three double bonds of carbon chain is named as a oleic, linoleic, and linolenic fatty acid components respectively [13]. The main limitations of vegetable oil are its poor low temperature behavior, oxidation and thermal stability and gumming effect [14, 15]. These stabilities and pour point behavior can be ameliorated by transesterification. Moreover the inferior flow property does not affect much in the tropical countries. Quinchia et al. [16] stated that, improving the potentiality of biolubricants some technical properties including available range of viscosities are need to improved. To do so, environmentally friendly viscosity modifier can be used. viscosity is the most important property for the lubricants since it determines the amount of friction that will be encountered between sliding surfaces and whether a thick enough film can be build up to avoid wear from solid-to-solid contact. Since little chance of viscosity with fluctuations in temperature is desirable to keep variations in friction at a minimum, fluid often are rated in terms of viscosity index. The less the viscosity is changed by temperature, the higher the viscosity index. Ethylene–vinyl acetate (EVA) and styrene–butadiene–styrene (SBS) copolymers were used to increase the viscosity range of high-oleic sunflower oil, in order to design new environmentally friendly lubricant formulations with increased viscosities. The maximum kinematic viscosities, at 40 and 100 ◦C, were increased up to around 150–250 cSt and 26–36 cSt, respectively [17].

Despite of having lot of advantages of biolubricant over petroleum based lubricant, the attempt to formulate the biolubricant and its applications are very few. Thus, in this article we sought to extend our investigation and to test the tribological characteristics and compatibility of non-edible Jatropha oil based biolubricant for the automotive application. The reason of selecting Jatropha oil as a base stock is it does not contend with the food and can be grown in marginal land.

2. Experimental

2.1. Lubricant sample preparation

There were six different types of lubricant sample were investigated in this study. The lubricant SAE 40 was used as a base lubricant and comparison purpose. Others samples were prepared by mixing of 10%, 20%, 30%, 40% and 50% Jatropha oil in SAE 40. The samples were mixed with the base lubricant by a homogeneous mixture machine.

2.2. Friction and wear evaluation

The apparatus used in the friction and wear testing process were Cygnus Friction and Wear Testing Machine which is connected with a personal computer (PC) with data acquisition system. It is a tri-pin-on-disc machine which is conducted by using three pins on a disc as testing specimens. Specifications of the Cygnus Test Machine are tabulated in Table 1. The block diagram of friction and wear testing are shown in Fig. 1. During the test the load of 30N and rotational speed of 2000 rpm were applied on pin.

Parameter	Value
Test Disc Diameter	110.0 mm
Test Pin Diameter	6.0 mm
Test Disc Speed Range	25 to 3000 rpm
Motor	Tuscan; (2000 rpm, 1.5 kW)
Load Range	0 KG to 30 KG
Electrical Input	220 Volt AC 50 Hz

Table 1. Specification of Cygnus wear testing machine

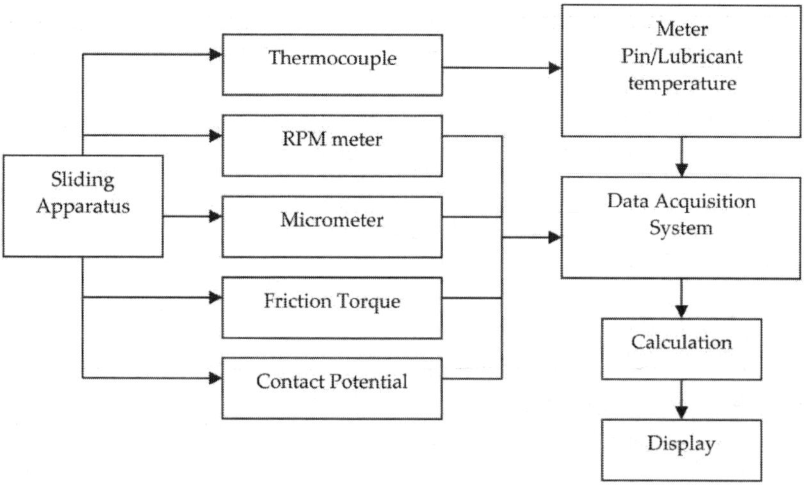

Figure 1. Block Diagrams of Friction and Wear Testing

2.3. Preparation of the specimen

The specimens were prepared from aluminum and cast iron material. Aluminum was used to build three pin and cast iron is used for disc specimen. The construction geometry and the dimension are shown in Fig. 2. Prior to conduct the test it was ensured that the surface of the specimens are cleaned properly i. e, free from dirt and debris. Alcohol was used for cleaning purpose.

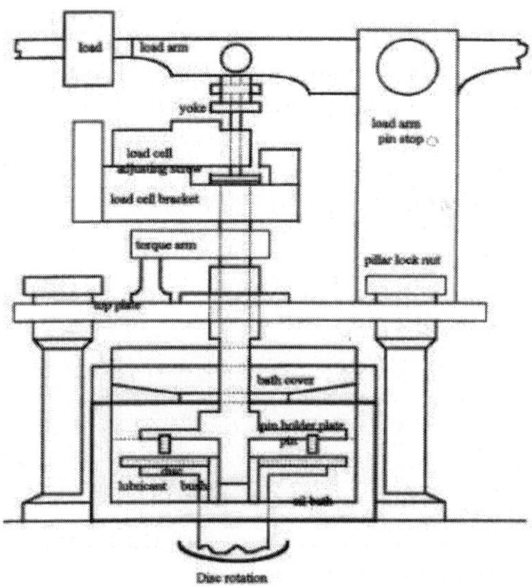

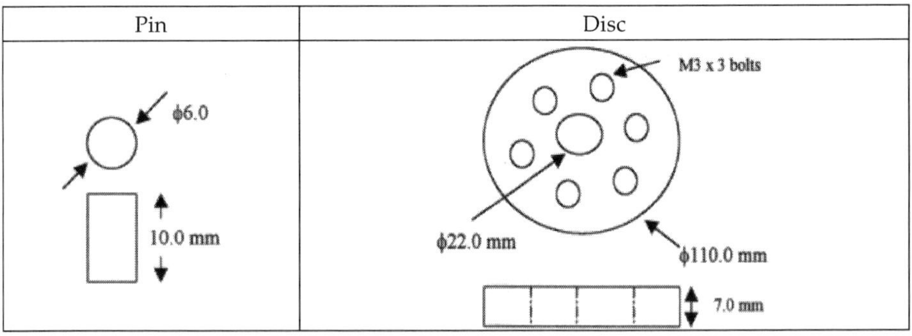

Figure 2. Schematic diagram of the experimental set up and dimensions geometry's of pins and disc specimen

2.4. Lubricant analyses

Multi element oil analyzer (MOA) was used to measure the wear elements in the lubricants by Atomic Emission Spectroscopy (AES). Whereas, for viscosity measurement the automatic Anton Paar viscosity meter was used with standard ASTM D 445. Viscosity was measured for both 40°C and 100°C controlled bath temperatures.

3. Results and discussion

3.1. Friction and wear characterization

Fig. 3 show the pins wear as a function of sliding time for various Jatropha oil blended biolubricants. At the operating condition of 2000 rpm and 30 N loads, the linear pin wear varied from 0.02 to 0.05 mm. It is observed that the maximum wear occurred in the beginning of the experiment using biolubricants. It can be seen form the Fig. 3, that the maximum wear was occurred for JBL40 while the minimum wear was observed for JBL10. The results can be attributed to the maximum ability of the JBL 10 biolubricant film to protect metal to metal contact and keep consistency throughout the operation time while this ability is least for JBL40. It can also be seen that the rate of wear throughout the time is almost identical for the biolubricants whereas, the reducing trend is observed for the base lubricant. At the beginning of the test, the wear rate was very fast for few minutes which are known running-in period. During this period, the asperities of the sliding surface are cut off and the contact area of the sliding surface grows to an equilibrium size. After certain period of time, equilibrium wear condition between pins and disc surface was established and thereby the wear rate became steady. It can be identified from the Fig. 3 that the biolubricants JBL 30, JBL 40 and JBL 50 showed high wear while base lubricant, JBL 10 and JBL 20 impart low pin wear and their values are nearly same with each other.

Fig. 4 sows the loos of material from the pin for different percentage of biolubricant samples. It seems quite clear that the loos of material from the pins are highest for 50% biolubricant and that is least for base lubricant. It can also be interpreted that the loos of material from

JBL 10 is almost similar with base lubricant and this loos of material is increasing with increasing biolubricant percentages.

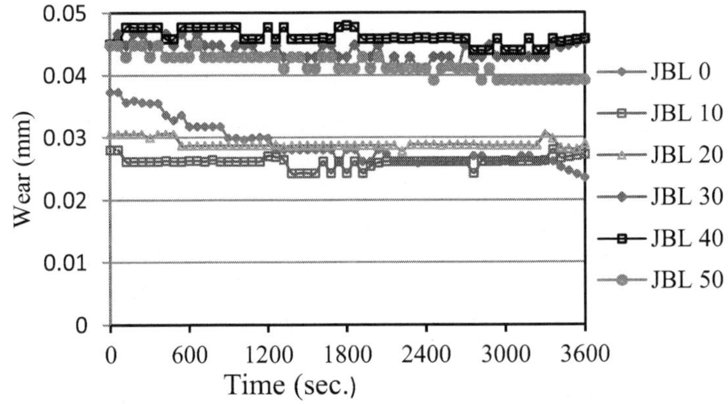

Figure 3. The linear pin wear as a function of sliding time for various Jatropha oil biolubricants.

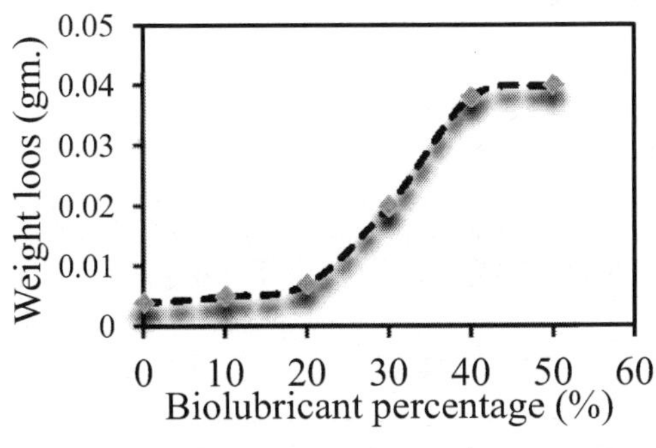

Figure 4. Loos of material form the pin for various biolubricant percentages

3.2. Coefficient of friction

Fig.5 shows the friction coefficient plotted against the sliding time for various Jatropha oil biolubricants. The results of the figure depict that the lubricant regime that occurred during the experiment were the boundary lubrication with the value of friction coefficient for boundary lubricant in the range of 0.001 to 0.2 except for 50% of Jatropha oil biolubricant. For JBL 0, it can be seen that the coefficient of friction is highest at the beginning and then it fell down rapidly and became least with compared to all tested samples after half of the operation time. The biolubricant percentage from 10 to 40% showed likely to be similar coefficient of friction (μ) which is almost 0.15. Whereas, the 50 % added Jatropha oil showed the coefficient of friction value of ~ 0.225 throughout the operation time. The fatty acid component of biolubricants formed multi and mono layer on the surface of the rubbing zone and make stable film to prevent the contact between the surfaces.

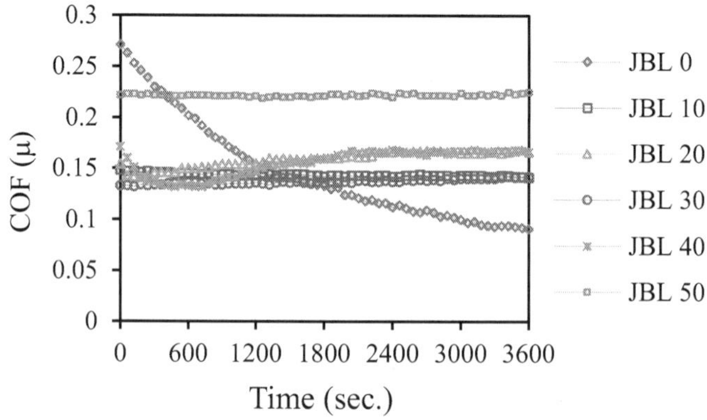

Figure 5. The Coefficient of friction as a function of sliding time for various Jatropha oil biolubricants

3.3. Lubricants temperature

Fig.6 shows the relationship of the averages oil temperature of varies percentage of Jatropha oil biolubricants with the sliding time. The rise of temperature during the running hour (1 h) for JBL 10 is least while the highest change is occurred for JBL 40 which is 11.77°c and 25.49°C respectively. The temperature rises of other samples are of 12.8°C, 18.65°C and 13. 66°C for 20% 30% and 50% Jatropha oil added biolubricants respectively. The results of the Fig. 6 show that the JBL 10 has the highest potentiality to retain its property without much changing its temperature. From the figure it can also be interpreted that up to 30 minutes rate of change

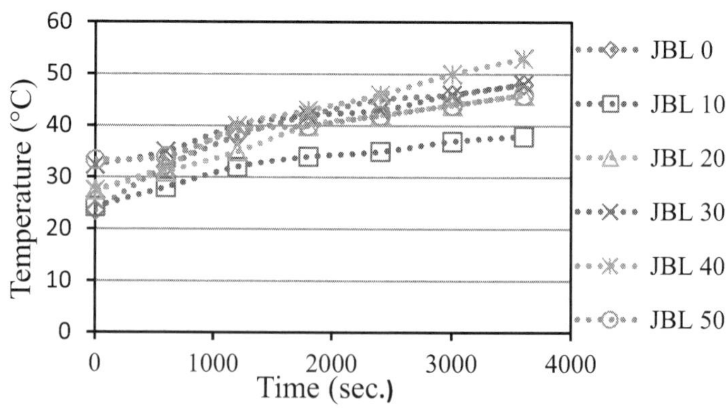

Figure 6. The Lubricant Temperature as a function of sliding time for various Jatropha oil biolubricants

of temperature is high while the changing rate is low for second half of the operation time. It can be explained that during second half of the operation time heat produced in the lubricant due friction and the heat dissipated to the outside is nearly equilibrium.

3.4. Viscosity

Viscosity is the measure of resistance to flow [18]. Table 2 shows the viscosity grade requirement for the lubricants set by International standard organization (ISO), while Fig. 7 shows the viscosity of tested different biolubricant samples. The comparison of the results of the Fig.7 with that of ISO grade illustrates that in case of 40°C, the biolubricants JBL 40 and JBL 50 did not meet the ISO VG100 requirement. On the other hand all other biolubricants meet the entire ISO grade requirement as well. It can also be noted that the viscosity of biolubricants are much higher than standard requirements

Kinematic viscosity	ISO VG32	ISO VG46	ISO VG68	ISO VG100
@ 40°C	>28.8	>41.4	>61.4	>90
@ 100°C	>4.1	>4.1	>4.1	>4.1

Table 2. ISO Viscosity grade requirement [19]

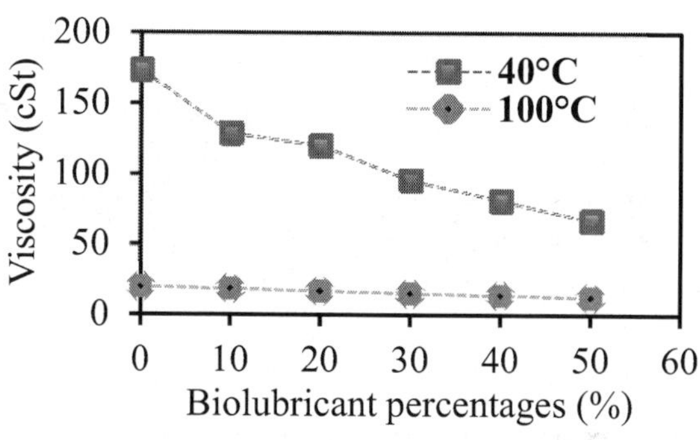

Figure 7. The viscosity of various percentages of biolubricants at 40°C and 100°C

3.5. Elemental analysis

The aim of the elemental analysis by using Multi Element Oil Analyzer (MOA) is to determine the kinds and amount of metal contain in the lubricating oil. Table 3 shows the elemental analysis of tested lubricant sample by using MOA before and after the test. From the Table 3, it can be noticed that the base lubricant contains higher Silver (Ag), Zinc (Zn), Phosphorus (P), Magnesium (Mg) and Boron (B) with in high percentage compared to other element while, in pure Jatropha oil, Calcium (Ca) and Silicon (Si) are the higher element compared with other element. Some of the elements are used as additive in the lubricant to ameliorate the lubricants tribological properties. From the results, increasing number of iron

Parameters	Types of Lubricant												
	JBL 0		JBL 10		JBL 20		JBL 30		JBL 40		JBL 50		Jatropha oil
Test	Before	After	Before	After	Before	After	Before	After	Before	After	Before	After	
Iron (Fe)	0.00	2.00	1.00	2.00	1.00	3.00	1.00	3.00	1.00	6.00	2.00	6.00	2
Aluminum (Al)	0.00	15.00	0.00	81.00	0.00	188.00	0.00	205	0.00	211.0	0.00	76.00	0
Copper (Cu)	0	1.00	0.00	3.00	1.00	1.00	1.00	1.00	1.00	7.00	2.00	5.00	3
Lead (Pb)	3	4.00	4.00	5.00	2.00	4.00	3.00	4.00	3.00	3.00	3.00	2.00	0
Tin (Sn)	0.00	0.00	0.00	0.00	0.00	0.00	0.00	0.00	1.00	0.00	2.00	2.00	4.5
Nickel (Ni)	2.00	2.00	3.00	3.00	1.00	3.00	3.00	3.00	3.00	3.00	2.00	2.00	1.5
Titanium (Ti)	0.00	0.00	1.00	1.00	0.00	1.00	0.00	1.00	0.00	0.00	1.00	1.00	1
Silver (Ag)	108	103	0.00	0.00	0.00	0.00	0.00	0.00	0.00	0.00	0.00	0.00	0
Molybdenum (Mo)	3.00	3.0	4.00	6.00	2.00	3.00	4.00	3.00	4.00	6.00	3.00	4.00	1.5
Zinc (Zn)	1000	771	903	716	1000	829.0	911.0	851.0	942.00	900.0	946.00	832.00	1
Phosphorus (P)	500.00	428	471	441	462.00	440.0	435.00	408.0	387.00	394.0	348.00	294.00	45
Calcium (Ca)	18.00	17.00	21.00	29.00	23.00	21.0	28.00	27.0	35.00	33.00	37.00	30.00	40
Magnesium (Mg)	748.00	637.0	572.	616.00	557.00	435.0	503.00	527.0	508.00	483.0	409.00	211.00	27
Silicon (Si)	5.00	4.00	6.00	10.00	6.00	15.0	8.00	12.0	9.00	13.0	14.00	7.00	16
Sodium (Na)	2.00	1.00	2.00	5.00	2.00	2.0	3.00	4.00	5.00	4.00	3.00	4.00	4
Boron (B)	60.00	54.00	52.00	58.00	52.00	28.0	52.00	32.0	44.00	44.0	40.00	21.00	0.5
Vanadium (V)	0.00	1.00	0.00	1.00	0.00	0.00	1.00	0.00	1.00	0.00	1.00	1.00	1

*All values are in ppm

Table 3. Elemental analysis of tested lubricant sample

(Fe) and aluminum (Al) molecules are observed with increasing percentages of Jatropha oil in the base lubricants. The source of Fe and Al are mainly cast iron plate and aluminum plate. Due to lower hardness of the aluminum pin the extraction of aluminum molecule form the pin is much higher than cast iron plate. The changes of other elements were observed before and after the test. It is clear from the elemental analysis that, most of elements were decreased after the test, by oxidizing and the chemical interaction among the elements.

3.6. Surface texture analysis

There are various types of wear in the mechanical system, such that abrasive wear, adhesive wear, fatigue wear and corrosive wear. Since the lubricant regime occurred in this experiment was boundary lubrication thereby, abrasive wear, adhesive wear, fatigue wear and corrosive wear were observed in to the rubbing zone. All these wears mechanisms found in this experiments but the mostly the wear phenomenon were abrasive and adhesive wear. This is because of an existence of straight grooves in the direction of the sliding direction. These grooves exist because the asperities on the hard surface (disc) touched the soft surface (pins) and hade a close relationship with the thickness of lubrication film. The optical images of the tested cast iron plate using various types of biolubricants are shown in Fig. 8. Referring to the Fig. 8, it is found that the wear increases with increasing percentage of Jatropha oil in the biolubricants. Reduction of lubricant film thickness leads to the surfaces to come closer to each other and cause higher wear.

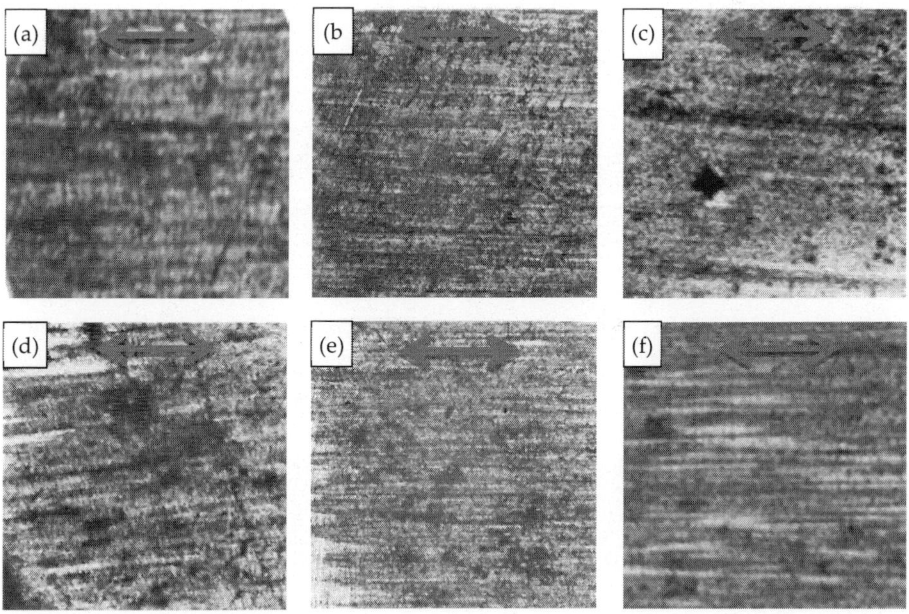

Figure 8. Optical image of the surface of the cast iron plate for different biolubricants (magnification 30 ×): (a): JBL 0, (b): JBL10, (c): JBL 20, (d): JBL 30, (e): JBL 40, (f): JBL 50

4. Conclusions

Based on the experimental study the following conclusion can be drawn:

1. The rates of wear for various percentage of biolubricant were different. Moreover the wear rate for 10% Jatropha added biolubricants were almost identical with base lubricant.
2. Lower the resistance to wear, higher coefficient of friction.
3. At the beginning of the test rate of wear as well as rise in temperature were high. With respect to wear rate and rise in temperature during entire operation time, the JBL 10 biolubricant showed best performance in terms of its ability to withstand its properties.
4. From the elemental analysis of the biolubricants, it is found, Iron and Aluminum were increased after the test due to the loos of material from the pin and the disc while, some element like Phosphorus, Calcium and Magnesium were decreased by oxidizing and due to other chemical interaction.
5. In terms of viscosity, almost all biolubricants met the ISO viscosity grade requirement whereas, 40% and 50% addition of Jatropha oil do not meet the ISO VG 100 requirement at 40°C.

According to the experimental result, it can be recommended that the addition of 10% Jatropha oil in the base lubricant is the optimum for the automotive application as it showed best overall performance in terms of wear, coefficient of friction, viscosity, rise in temperature etc.

Author details

M. Shahabuddin*, H.H. Masjuki and M.A. Kalam
Centre for Energy Sciences, Faculty of Engineering, University of Malaya, Kuala Lumpur, Malaysia

Acknowledgement

The authors would like to acknowledge the Department of Mechanical Engineering, University of Malaya, Ministry of Higher Education (MOHE) of Malaysia for HIR grant (Grant No. UM.C/HIR/MOHE/ENG/07) and ERGS grant no ER022-2011A which made this study possible.

5. References

[1] Salih N, Salimon J, Yousif E. Synthetic biolubricant basestocks based on environmentally friendly raw materials. Journal of King Saud University –Science 2011.

---

* Corresponding Author

[2] Adhvaryu A, Liu Z, Erhan S. Synthesis of novel alkoxylated triacylglycerols and their lubricant base oil properties. Industrial Crops and Products 2005; 21:113-119.

[3] Shahabuddin M, Masjuki HH, Kalam MA *et al.* Effect of Additive on Performance of C.I. Engine Fuelled with Bio Diesel. Energy Procedia 2012; 14:1624-1629.

[4] Siniawski MT, Saniei N, Adhikari B, Doezema LA. Influence of fatty acid composition on the tribological performance of two vegetable-based lubricants. Journal of Synthetic Lubrication 2007; 24:101-110.

[5] Salunkhe DK. World oilseeds: chemistry, technology, and utilization. 1992.

[6] Hwang HS, Erhan SZ. Lubricant base stocks from modified soybean oil. AOCS Press: Champaign, IL; 2002.

[7] Ing TC, Rafiq AKM, Syahrullail S. Friction Characteristic of Jatropha Oil using Fourball Tribotester. In: Regional Tribology Conference - RTC2011. Langkawi, Kedah, Malaysia: 2011.

[8] M. Shahabuddin, M. A. Kalam, H. H. Masjuki, M. Mofijur. Tribological characteristics of amine phosphate and octylated/butylated diphenylamine additives infused biolubricant. Energy Education Science and Technology Part A: Energy Science and Research 2012; 30:89-102.

[9] Totten G.E., Westbrook S.R, Shah R.J. Fuels and Lubricants Handbook: Technology,Properties, Performance, and Testing. 2003. 885–909. p.

[10] Liaquat AM, Masjuki HH, Kalam MA *et al.* Application of blend fuels in a diesel engine. Energy Procedia 2012; 14:1124-1133.

[11] Jayadas N, Nair KP. Coconut oil as base oil for industrial lubricants--evaluation and modification of thermal, oxidative and low temperature properties. Tribology international 2006; 39:873-878.

[12] Fox N, Stachowiak G. Vegetable oil-based lubricants—a review of oxidation. Tribology international 2007; 40:1035-1046.

[13] Waleska C, David EW, Kraipat C, Joseph MP. The effect of chemical structure of base fluids on antiwear effectiveness of additives. Tribol. Int. 2005; 38:321–6.

[14] Ponnekanti N, Kaul S. Development of ecofriendly/biodegradable lubricants: An overview. 2012.

[15] Mofijur M, Masjuki HH, Kalam MA *et al.* Palm Oil Methyl Ester and Its Emulsions Effect on Lubricant Performance and Engine Components Wear. Energy Procedia 2012; 14:1748-1753.

[16] Quinchia L, Delgado M, Valencia C *et al.* Viscosity modification of different vegetable oils with EVA copolymer for lubricant applications. Industrial Crops and Products 2010; 32:607-612.

[17] Quinchia L, Delgado M, Valencia C *et al.* Viscosity modification of high-oleic sunflower oil with polymeric additives for the design of new biolubricant formulations. Environmental science & technology 2009; 43:2060-2065.

[18] Shahabuddin M, Kalam MA, Masjuki HH *et al.* An experimental investigation into biodiesel stability by means of oxidation and property determination. Energy 2012.

[19] Rudnick LR. Automotives Gear Lubricants, Synthetics, mineral oils, and bio-based lubricants: chemistry and technology. Taylor and Francis, Florida; 2006.

Section 3

Friction

Chapter 9

Friction in Automotive Engines

H. Allmaier, C. Priestner, D.E. Sander and F.M. Reich

Additional information is available at the end of the chapter

http://dx.doi.org/10.5772/51568

1. Introduction

The current situation of the automotive industry is a challenging one. On one hand, the ongoing trend to more luxury cars brings more and more benefits to the customer and is certainly also an important selling point. The same applies to the increased safety levels modern cars have to provide. However, both of these benefits come with a severe inherent drawback and that is extra weight and, consequently, higher fuel consumption. On the other hand, increased fuel consumption is not only a disadvantage due to the ever rising fuel costs and the corresponding customer demand for more efficient cars. Due to the corresponding greenhouse gas emissions it is also in the focus of the legislation in many countries. Commonly road transport is estimated [13, 15] to cause about 75-89 % of the total CO_2 emissions within the world's transportation sector and for about 20% of the global primary energy consumption [12]. These values do not stay constant; in the time from 1990 to 2005, the required energy for transportation increased by 37% [11] and further increases are expected due to the evolving markets in the developing countries. As industrial emissions decrease, the rising energy demand in the transport sector is expected to be the major problem to achieve a significant greenhouse gas reduction [26]. Consequently, about all major automotive markets introduce increasingly strict emission limits like the national fuel economy program implemented in the CAFE regulations in the US, the EURO regulation in the European Union or the FES in China.

In particular, the European union introduced a limit for the average CO_2 emissions for all cars to be available on the European market of 130 g CO_2/km by 2015[1]. Further, a long term target of 95 g CO_2/km was specified for 2020 [6]. To put this into perspective, the average fleet consumption in 2007 was 158 g CO_2/km and it had taken already about 10 years to get down to this value from the 180 g CO_2/km that were achieved in 1998. Now a larger reduction is required in less time. The required reduction of emissions brings also a direct benefit for the customer as the fuel consumption is lowered. It is estimated [11] that the

[1] The limit applies to the average fleet consumption of every car manufacturer, calculated by averaging the fuel consumption of all offered car models and weighted by the number of sold units of the specific models.

average car consumes every year about 169 litres of fuel only to overcome mechanical friction in the powertrain. There exist several efficient measures to lower the emissions, notably to decrease the weight of the car itself to reduce the energy needed for acceleration, to optimize the combustion process and, of course, to reduce all inherent power losses like the tyre rolling resistance, aerodynamic drag and mechanical losses of the powertrain (combination of engine and transmission) itself. These measures are, however, not straightforward as some have drawbacks or are even in conflict with each other. For example, smaller cars that offer reduced weight have commonly worse aerodynamic drag as their shape is more cube-like for practical reasons [8]. Also, a reduction in aerodynamic drag brings only a small benefit in the urban traffic due to the low cruising speeds involved. A reduction of tyre rolling resistance is hard to achieve without reduced performance in other areas like handling and traction [10]. Weight reduction is expensive as more and more lightweight and expensive materials have to be used, some of which also require a lot of energy in the production process. In addition, it was shown [8] that there is no positive synergy effect: the combination of several of the mentioned measures reduces their individual efficiency, such that their combination brings less benefit than anticipated.

In contrast, making the powertrain more efficient yields a proportional reduction in CO_2 emissions [8]. For low load operating conditions of Diesel engines friction reduction is even the prime measure to further decrease fuel consumption [16]. While currently a lot of work is done in the automotive industry to reduce the losses caused by the auxiliary systems like the oil or coolant pump, it was shown at hand of a specific engine that the potential for friction reduction in the ICE itself is of comparable magnitude [16].

2. Sources of friction in ICEs

Before any efficient measures to reduce friction in engines can take place, the main friction sources need to be known. At the Virtual Vehicle Competence Center, we use our friction test-rig as shown in Fig. 1 to investigate the sources of friction for a typical four cylinder gasoline engine; exemplary results for this engine are shown in Fig. 2.

The chart confirms the commonly propagated main sources of friction: the piston-liner contact is the cause for about 60% of the total mechanical losses, while the journal bearings in the crank train (main and big end bearings) contribute about 30%. Finally, the valve train generally represents the third main source of friction and typically causes losses that equal roughly about the half of the power losses in the journal bearings [19] (not included in Fig. 2).

While not only the amount of friction is different between the various sources, also the character of friction, namely the lubrication regime itself, is also notably different. While the journal bearings are generally full film lubricated with no metal metal contact occurring under normal operating conditions, parts of the piston assembly experience metal-metal contact under high load. In particular, the top ring of the piston has metal-metal contact every time it passes the top dead center as no oil can reach this point. This is of particular severity as at firing top dead center a large force acts on the top ring and presses it onto the cylinder liner during the downward motion of the piston. Besides the fact that the piston assembly has generally been the largest contributor to the total mechanical losses, it has several other important functions. Amongst others it has to seal the combustion chamber in both directions,

Figure 1. Friction measurement test-rig FRIDA during build-up at the Virtual Vehicle Competence Center. It is shown being applied to an inline four cylinder gasoline engine with 1.8 litres total displacement.

both to avoid so-called blow-by gases from entering the engine housing, as well as to control the amount of lubricant being left on the cylinder liner. The blow-by gases have to be controlled as these both cause a loss in convertible energy by decreasing the available cylinder pressure as well as have a negative deteriorating impact on the lubricant properties. The oil being left on the cylinder liner needs to be carefully controlled as well: while a certain amount of oil is necessary to provide sufficient lubrication for the piston rings, it is burned during combustion. Burning too much oil needs to be avoided not only for practical reasons as it needs to be replaced (increased service demand), but also as some of its contents are problematic for the exhaust aftertreatment systems.

Additionally, depending on operating condition unstable behaviour of the piston rings may occur [27] like ring flutter (rapid oscillating movement of the piston ring in its groove) or ring collapse (inward forces on the ring exceed the ring tension), which needs to be avoided in practical designs. To summarize, the piston assembly has to fulfil many functions. For focusing solely on friction it is, therefore, not used in this work. In the following, the second largest contributor to the total losses in engines, namely the journal bearings, are discussed.

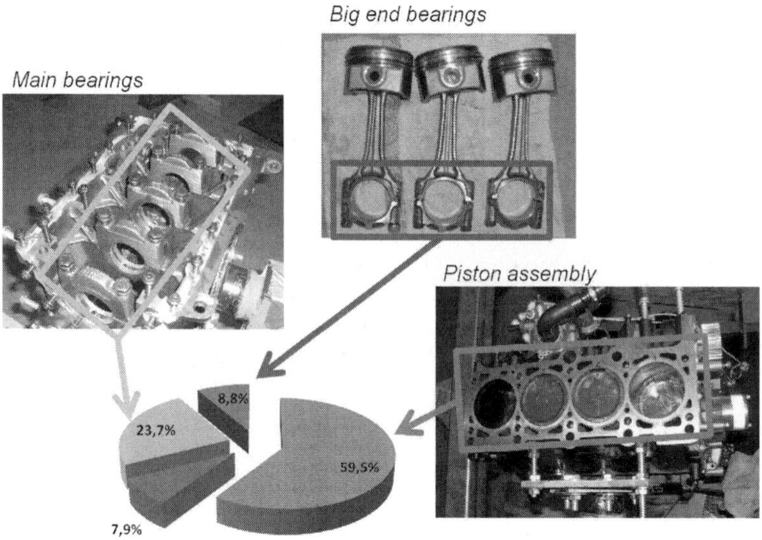

Figure 2. Examplary relative contributions to the total friction losses in an inline 4 cylinder gasoline engine with 1.8 litre total displacement. Shown in blue is the contribution of the piston assembly, in green the contribution of the main bearings, in violet the amount caused by the big end bearings and, finally, the red part shows the contribution of all other components like seals etc. The valve train is not included in these results, also all auxiliary systems (oil pump etc.) are removed.

3. Calculating power losses due to friction in the journal bearings

In contrast to the piston assembly that has to perform a large number of tasks which are partially conflicting as previously discussed, journal bearings are due to their apparent simplicity particularly suited to discuss the sources of friction.

Journal bearings are from their appearance simple devices; generally formed from sheet metal they are typically low cost parts, with one bearing shell costing a few single Euros or less. However, this simplicity is misleading, as in fact they have to combine a wide range of properties which impose conflicting requirements on the material properties to be used. While the bearing material should be hard to resist wear, in the engine it shall also embed well debris particles that originate from wear or even from the original manufacturing process of the engine housing. For the latter property softer materials are beneficial which conflicts with the requirement to resist wear. These requirements led to the development of multi layer bearings, where each layer is optimized for a specific task.

In the following a method is described that accounts for many of the essential physical processes that occur in journal bearings during operation and allows to accurately predict the power losses due to friction. The method is developed while discussing these processes and its validity is shown by numerous comparisons to experimental data.

While the focus in the following is on monograde oils as they are used in large stationary engines, the results also apply correspondingly to multigrade oils with their shear rate dependence taken into account.

In the following the results from a number of works are presented in a shortened form with a particular focus on the results and their context. All details can be found in the original works [1–3, 24, 25].

3.1. An isothermal EHD approach

In an ICE, journal bearings are generally exposed to different operation conditions in terms of load, speed and temperature. As depicted in Fig. 3, depending on relative speed, load and viscosity the operating conditions reflected as friction coefficient may range from purely hydrodynamic lubrication with a sufficiently thick oil film to mixed or even boundary lubrication with severe amounts of metal to metal contact.

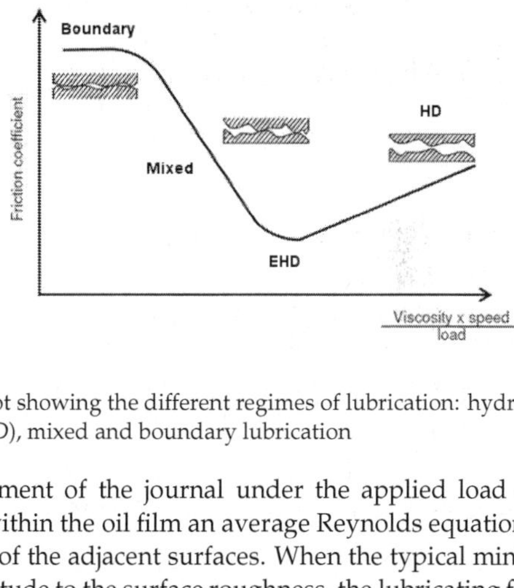

Figure 3. The Stribeck-plot showing the different regimes of lubrication: hydrodynamic (HD), elastohydrodynamic (EHD), mixed and boundary lubrication

To calculate the movement of the journal under the applied load and the corresponding pressure distribution within the oil film an average Reynolds equation is used, that takes into account the roughness of the adjacent surfaces. When the typical minimum oil film thickness is of comparable magnitude to the surface roughness, the lubricating fluid flow is also affected by the surface asperities and their orientation. To account for this modification of the fluid flow we use the average Reynolds equation as developed by Patir and Cheng [21, 22], which can be written in a bearing shell fixed coordinate system as

$$\begin{aligned} &-\frac{\partial}{\partial x}\left(\theta\phi_x\frac{h^3}{12\eta_p}\frac{\partial p}{\partial x}\right)-\frac{\partial}{\partial z}\left(\theta\phi_z\frac{h^3}{12\eta_p}\frac{\partial p}{\partial z}\right)+ \\ &+\frac{\partial}{\partial x}\left(\theta(\bar{h}+\sigma_s\phi_s)\frac{U}{2}\right)+\frac{\partial}{\partial t}\left(\theta\bar{h}\right)=0, \end{aligned} \tag{1}$$

where x, z denote the circumferential and axial directions, θ the oil filling factor and $h, \bar{h}$ the nominal and average oil film thickness, respectively. Further, U denotes the journal circumferential speed, η_p the pressure dependent oil viscosity and σ_s the combined (root mean square) surface roughness. ϕ_x, ϕ_z, ϕ_s represent the flow factors that actually take into account the influence of the surface roughness.

To describe mixed lubrication another process needs to be taken into account, namely the load carried by the surface asperities when metal-metal contact occurs.

The corresponding quantity is the asperity contact pressure p_{a} and together with the area experiencing metal-metal contact, A_{a}, and the boundary friction coefficient μ_{Bound} these yield the friction force R_{Bound} caused by asperity contact,

$$R_{\mathrm{Bound}} = \mu_{\mathrm{Bound}} \cdot p_{\mathrm{a}} \cdot A_{\mathrm{a}}. \tag{2}$$

To describe the metal-metal contact we use the Greenwood and Tripp approach [9], that is shortly outlined in the following.

The theory of Greenwood and Tripp is based on the contact of two nominally flat, random rough surfaces. The asperity contact pressure p_{a} is the product of the elastic factor K with a form function $F_{\frac{5}{2}}(H_s)$,

$$p_{\mathrm{a}} = KE^* F_{\frac{5}{2}}(H_s), \tag{3}$$

where H_s is a dimensionless clearance parameter, defined as $H_s = \frac{h-\bar{\delta}_s}{\sigma_s}$, with σ_s being the combined asperity summit roughness, which is calculated according to

$$\sigma_s = \sqrt{\sigma_{s,J}^2 + \sigma_{s,S}^2},$$

and $\bar{\delta}_s$ being the combined mean summit height, $\bar{\delta}_s = \bar{\delta}_{s,J} + \bar{\delta}_{s,S}$, where the additional subscript J and S denotes the corresponding quantities of the journal and the bearing shell, respectively. Further, E^* denotes the composite elastic modulus, $E^* = (\frac{1-\nu_1^2}{E_1} + \frac{1-\nu_2^2}{E_2})^{-1}$, where ν_i and E_i are the Poisson ratio and Young's modulus of the adjacent surfaces, respectively. The form function is defined as

$$\begin{aligned} F_{\frac{5}{2}}(H_s) =& 4.4086 \cdot 10^{-5} (4 - H_s)^{6.804} \quad \text{for} \quad H_s < 4 \\ =& 0 \quad \text{for } H_s \geq 4, \end{aligned} \tag{4}$$

which shows that friction due to asperity contact sets in only for $H_s < 4$ and further sensibly depends on the minimum oil film thickness as this quantity enters Eqn. 4 with almost 7th power.

For the calculation of the Greenwood/Tripp parameters a 2D-profilometer trace was used that was performed on an run-in part of the bearing shell along the axial direction.

Modern engine oils include friction modifying additives like zinc dialkyl dithiophosphate (ZDTP) or Molybdenum based compounds to lower friction and wear in case metal-metal contact occurs. For the Greenwood and Tripp contact model we employed in the following a boundary friction coefficient of $\mu_{\mathrm{Bound}} = 0.02$.

The different contributions to friction, as listed in Eqs. (1) and (3), are generally not independent from eachother. A reduction in lubricant viscosity, while decreasing hydrodynamic losses, may cause - depending on the load - an overly increase in asperity contact as the oil film thickness enters Eq. (4) with almost 7th power.

3.1.1. Testing Method

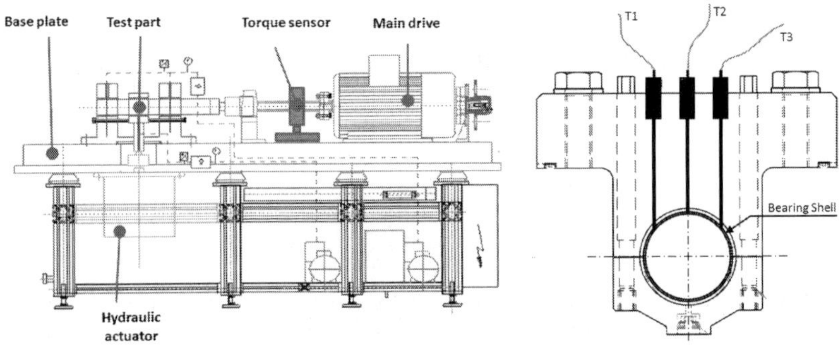

Figure 4. left: schematic drawing of the journal bearing test rig LP06: *test part* denotes the location of the test bearing, *torque sensor* the HBM T10F sensor used for friction moment measurement. Right: Drawing of the test con rod with test bearing showing the location of the temperatures sensors: T2 sits in the center at 0° circumferential angle, with T1 and T3 at ±45° circumferential angle, respectively.

MIBA[2]'s journal bearing test rig LP06 was used for the experimental measurements. It is sketched in Fig. 4 and consists of a heavy, elastically mounted base plate which carries the two support blocks, the test con rod with the hydraulic actuator and the driveshaft attached to the electric drive mechanism. The hydraulic actuator applies the load along the vertical direction, which is consequently defined as 0° circumferential angle.

The friction torques arising from all three journal bearings were measured at the driveshaft; for the comparisons load cycle averaged values of the friction moment (the load cycle is depicted in Fig. 5) are used.

The LP06 is equipped with a number of temperature sensors to capture the occurring temperatures at various points of the test rig; to this task temperature is measured by using thermocouple elements of type K that have an accuracy of ±1°C. Besides two temperatures in the con rod and the oil outflow temperature, the bearing shell temperatures of the test and support bearings are measured at three different points at the back of each corresponding bearing shell. As shown in Fig. 4, two of these temperature sensors are located at ±45° circumferential angle from the vertical axis and the third in the middle at 0° circumferential angle.

For the bearing tests following conditions were maintained: for test- and support-bearings steel-supported leaded bronce trimetal bearings with a sputter overlay were employed; for each test-run new bearings with an inner diameter of 76 mm and a width of 34 mm were used and mounted into the test rig with a nominal clearance of 0.04 mm ($1^0/_{00}$ relative clearance). A hydraulic attenuator applied the transient loads with the corresponding peak loads of either

[2] MIBA Bearing Group, Dr.-Mitterbauer-Str. 3 4663 Laakirchen, Austria

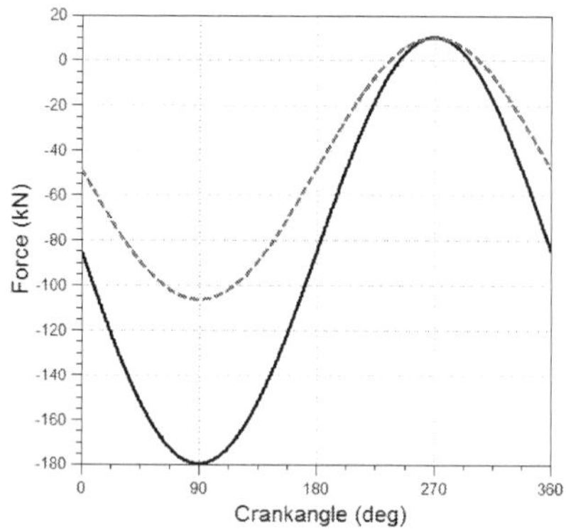

Figure 5. Plot of the loads applied to the test bearing: at a frequency of 50 Hz a sinusoidal load is applied along the vertical direction with a preload of -10kN and a peak load of either 180 kN for the 70 MPa load case (shown as solid black line) or a peak load of 106 kN for the 41 MPa load case (red dashed line).

41 MPa, 54 MPa, 70 MPa or 76 MPa. For a convenient comparison of the results to other works the peak load is expressed in MPa to account for the involved bearing dimensions. This is conducted by dividing the load force by the projected bearing area (product of bearing width and bearing diameter). Therefore, the peak loads of 106 kN and 180 kN correspond to 41 MPa or 70 MPa, respectively, for the present bearing dimensions (see also Fig. 5). In the following, the corresponding peak loads are used to distinguish between the different transient load cases.

The different oils were preconditioned to 80±5°C inflow temperature. After the test-run the wear at several points in the journal bearings was measured and the so obtained wear profiles were included in the simulation model.

3.1.2. *Simulation*

For the simulation a model of the LP06 was setup within an elastic multi-body dynamics solver (AVL-Excite Powerunit[3]). The simulation model consists of the test con rod including the test bearing, the two support-blocks with journal bearings and the shaft running freely, but supported by the adjacent bearings. All structure parts are modeled as dynamically condensed finite element (FE)-structures.

The three journal bearings, 76mm in diameter and 34mm width, are represented as EHD or TEHD-joints, respectively.

To obtain realistic dynamic lubricant viscosities for the calculations, the viscosities and densities of fresh SAE10/SAE20/SAE30 and SAE40 monograde oils were measured at different temperatures in the OMV-laboratory[4]. To obtain a pressure dependent oil-model for the simulation, the pressure dependency was impressed onto the measured viscosities by

[3] AVL List GmbH, Advanced Simulation Technology, Hans-List-Platz 1, 8020 Graz, Austria

[4] OMV Refining & Marketing GmbH, Uferstrasse 8, 1220 Wien, Austria

applying the well known Barus-equation [5] with the coefficients from [5]. The so resulting dynamic viscosities correspond qualitatively to experimental data [4]. Further, a dependence on hydrodynamic pressure was impressed onto the lubricant density following the data found experimentally by Bair et al. [4].

The dynamic viscosities and oil-densities are shown for the SAE10, SAE20, SAE30 and SAE40-oils in Fig. 6. As can be seen in these figures, a hydrodynamic pressure of about 60 MPa leads to roughly a doubling of the dynamic viscosity and, therefore, to a strong increase in the related hydrodynamic losses. While for now the presented calculations do not take into account the local temperatures of the lubricant in the bearing itself, the strong variation of the physical properties of the oil with temperature show the importance of defining a representative global lubricant temperature as discussed in the next subsection.

3.1.3. *Deriving the oil-temperature*

A plausible choice of this temperature is important as it directly relates to the lubricant viscosity and consequently acts on the minimum oil film thickness and the amount of asperity contact.

For the presented pressure dependent lubricant model, the calculation of the global oil temperature is straightforward: as the oil viscosity increases strongly for hydrodynamic pressures exceeding about 1 MPa, the hydrodynamic losses in the lubricant are expected to be dominated by this thickening in the high-load area of the bearing. This argument is also supported later on by the simulation results which predict hydrodynamic pressures of up to 120 MPa in large areas in the bearing. Following this line of argument, the global oil temperature is estimated from the measured bearing back temperatures, by averaging the test and support bearing back temperatures that are located at $\pm 45°$ circumferential angle (T_1, T_3) of the one in the high load zone, T_2, as shown in Fig. 4. Although the so obtained temperature is rather high in comparison to the oil inflow temperature it is expected to realistically estimate the hydrodynamic losses as well as the amount of asperity contact, as this temperature describes closely the oil viscosity in the high load zone.

The such calculated oil-temperatures are depicted in Table 1 for the load cases studied in the following and for simplicity the same oil temperature is used for all three bearings.

3.1.4. *Surface profiles*

For a sufficiently accurate calculation of the asperity contact, it is necessary to use realistic surface shapes in the simulation [24]. Ideal geometric shapes are not suitable for this task, as due to elastic deformation of the structure under load, the bearing pin would express overly large pressures on the outermost nodes of the bearing shell, leading to unrealistically high amounts of asperity contact. This in turn causes an overestimation of the friction moment.

Therefore, the bearing shell surface of the test bearing was measured for wear at several points after the test runs; the procedure is discussed in more detail in [24]. The wear data obtained from the two performed SAE10-oil test-runs were averaged and symmetrized as we do not include misalignments due to imperfect mounting in the simulation. The such obtained wear

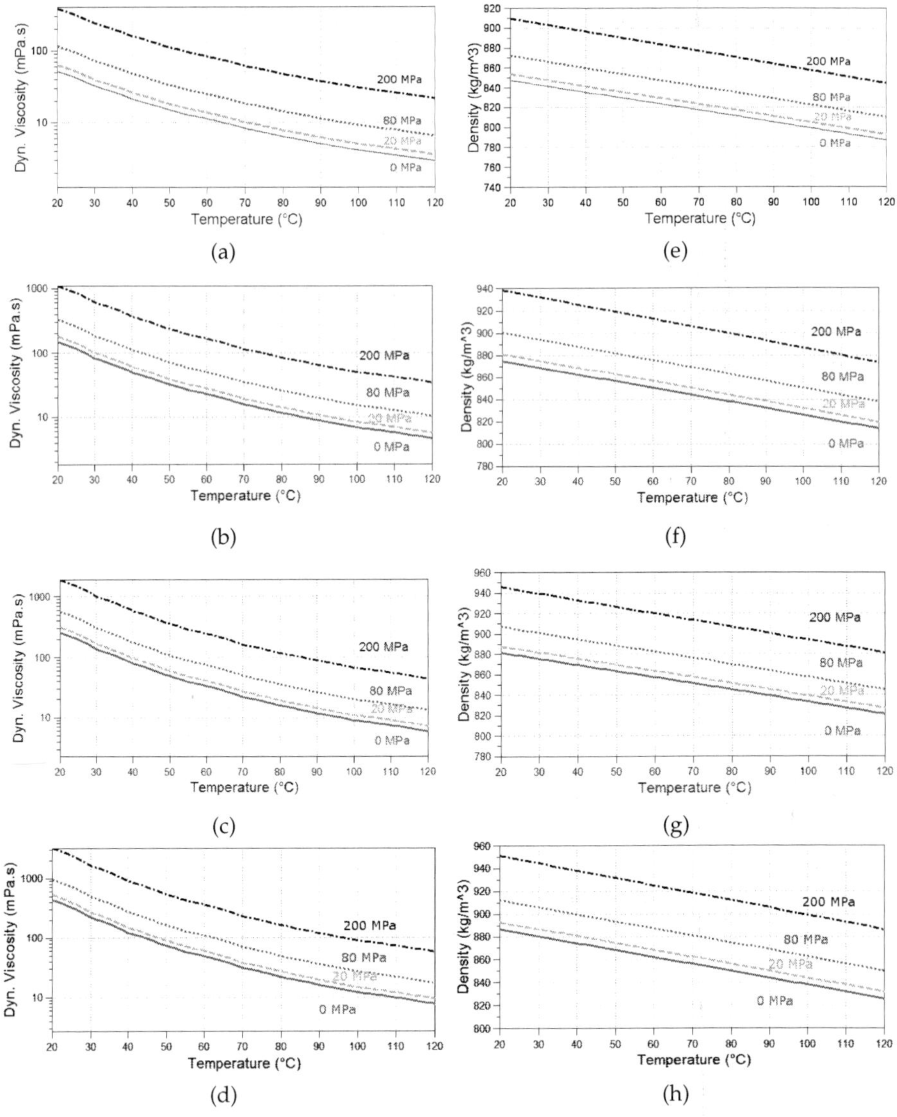

Figure 6. Dynamic viscosities (a-d) and densities (e-h) of the lubricants as described by the oil model for SAE10/SAE20/SAE30/SAE40 (top to bottom): the solid red lines denote the corresponding physical property at 0 Pa, the green dashed lines at 20 MPa, the blue dotted lines for 80 MPa and the black dash-dotted lines for 200 MPa.

SAE10	T_{41MPa} [°C]	T_{70MPa} [°C]
	87.4	94.6
SAE20	T_{41MPa} [°C]	T_{70MPa} [°C]
	89.2	96.8
SAE30	T_{41MPa} [°C]	T_{70MPa} [°C]
	89.8	97.9
SAE40	T_{41MPa} [°C]	T_{70MPa} [°C]
	91.8	99.6

Table 1. Calculated oil-temperatures (see text) for the studied load cases (denoted as subscript) and the different oils.

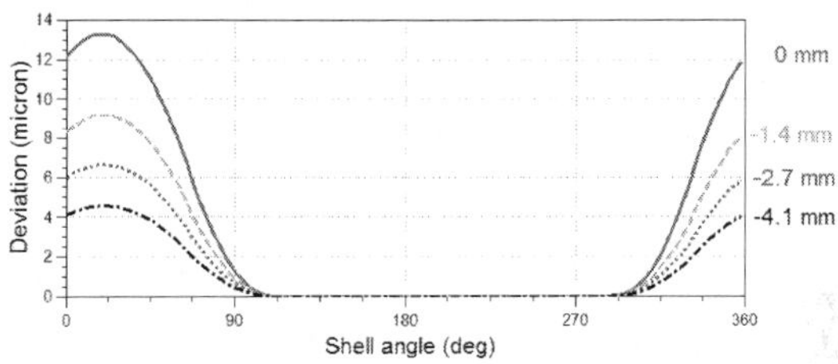

Figure 7. Surface profile used for the test bearing in the simulation, shown as deviation (in μm) from the nominal geometrical shape of the bearing shell for different axial cuts: the red line denotes the deviation at the outermost bearing nodes, 0mm from the shell edges, the green dashed line the deviation −1.4mm from the shell edges, the blue dotted line the deviation −2.7mm from the bearing edges and the black dash-dotted line shows the deviation −4.1mm from the bearing edges. The deviations in the middle of the journal bearing, along axial direction from −4.1mm to −17mm from the bearing edges, stay rather constant and are almost identical to the deviation for −4.1mm away from the bearing edges.

profiles, depicted in Fig. 7, were used for all consequent calculations (all lubricants and loads) as it was found that the results with individual wear profiles prepared for every lubricant caused only negligible differences in comparison to the results obtained with the SAE10-wear profile. Similar wear profiles were used for the support bearings, which, however, are less critical as each of these carries only half of the total load.

3.1.5. Results

In the following, the results obtained from simulation are compared to the experimental results. For this task, the results are discussed starting from full fluid film lubrication (purely hydrodynamic losses), as it is the case for SAE40, to working conditions which progress increasingly into mixed lubrication, like it is the case for SAE10, where friction power losses due to metal-metal contact become significant.

The simulations for the different lubricants were conducted and the calculated friction moments are shown and compared to the experimental values in Fig. 8. The obtained experimental values range from as low as 3 to 8 Nm, where the individual average friction torques are measured (with the exception of the 3000rpm cases where only a smaller amount of measurement data is available) with an accuracy of about 0.5 Nm (denoted as black bars in the plot).

Regarding the general trends, the resulting average friction moment scales with the applied load and for a given load, with the journal speed. For a specific load/journal speed case, the friction moment depends significantly on the different lubricant viscosities.

Due to the large number of cases shown in Fig. 8 the focus will be in particular on the results for a journal speed of 2000rpm and two different load cases in the following.

Comparing the results predicted by simulation to the experimental data, we start by looking first at the SAE40 lubricant. From experience it is known that for a load of 41 MPa SAE40 oil provides sufficient lubrication for reliable long term use and, thus, avoids asperity contact. This is confirmed by the results from simulation, as these do not show any asperity contact, see Fig. 9. The same is almost true for the 70 MPa case, although simulation predicts about 0.5 W power losses due to the metal-metal contact in this case. However, this amount is negligible compared to the total losses of 626 W.

Starting from the purely hydrodynamic case with SAE40 lubrication, excluding one case all results from simulation are within the measurement uncertainty of the experimental results and commonly even closer to the average friction torque. The result calculated for the excluded case is also close to the measurement uncertainty.

Lowering the lubricant viscosity by about 25% compared to the SAE40 lubricant, one arrives at the cases with lubrication with SAE30. Here, the onset of mixed lubrication occurs for the highest dynamic load of 70 MPa at the lowest journal speed of 2000rpm. For this smaller load, there is still no asperity contact predicted and all losses are caused by the dynamic viscosity of the lubricant. Beginning with this lubricant class the validity of the chosen asperity contact model starts to get tested. While the power losses are still strongly dominated by hydrodynamic losses, the power losses due asperity contact contribute about 1% (7 W averaged over a full crank cycle) to the total power losses. The corresponding asperity contact pressure (shown in Fig. 10 for all lubricants for the load case of 70 MPa and a journal speed of 2000rpm), however, increases for the SAE30 lubricant to a maximum of 15 MPa, which indicates that wear starts to occur in the highly loaded parts of the journal bearing.

However, comparing the results from simulation with the experimental data, the same level of accuracy can be seen in Fig. 8. In fact, the actual predicted value from simulation is generally even closer to the measured averaged value.

Lowering the lubricant viscosity once more to SAE20, the journal bearing experiences more and more mixed lubrication. While metal-metal contact is still absent for a load of 41 MPa, it starts to become significant for the 70 MPa load case. In total the losses due to metal-metal contact are still rather low representing only about 4% of the total power losses for the worst case as shown in Fig. 9. Due to the reduced viscosity in comparison to SAE30-oil, the experimentally observed average friction moments are reduced, however, asperity contact

(a)

(b)

(c)

(d)

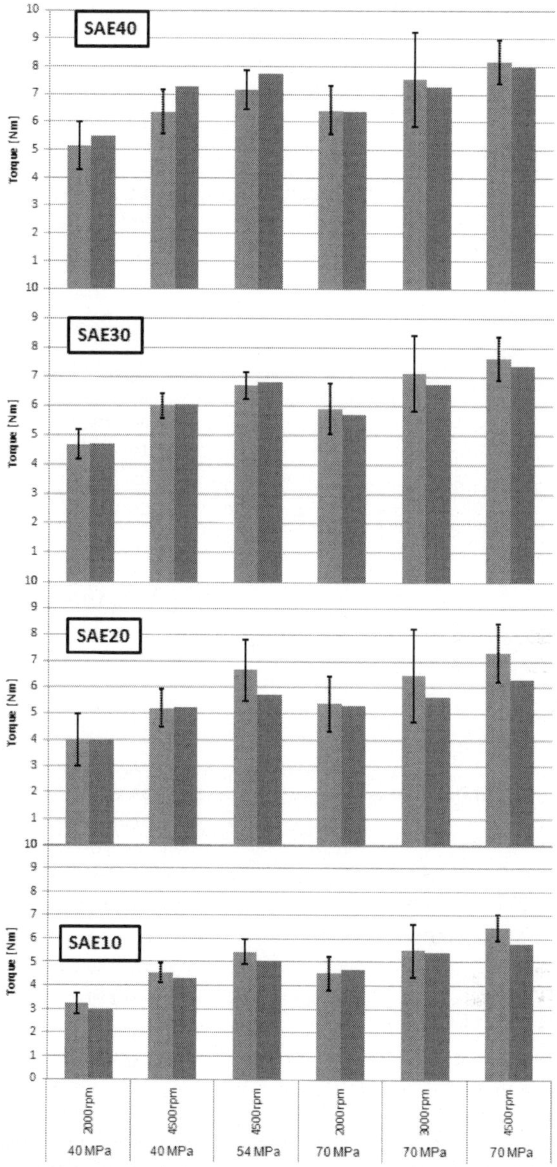

Figure 8. Comparison of the friction torques measured on the test-rig (blue) together with the measurement uncertainty (black bars) with the results from simulation (illustrated in red) for different journal speeds (2000/3000/4500rpm), different dynamic loads (40/54/70 MPa) and different lubricants (SAE40/SAE30/SAE20/SAE10).

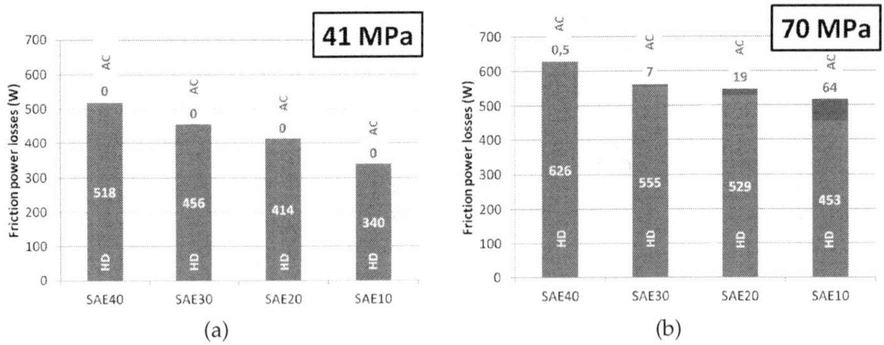

Figure 9. Comparison of the contributions to the friction power losses in the test bearing for a load of (a) 41 MPa and (b) 70 MPa for 2000rpm journal speed calculated for the different oils: hydrodynamic losses are denoted as HD and losses due to asperity contact are denoted as AC.

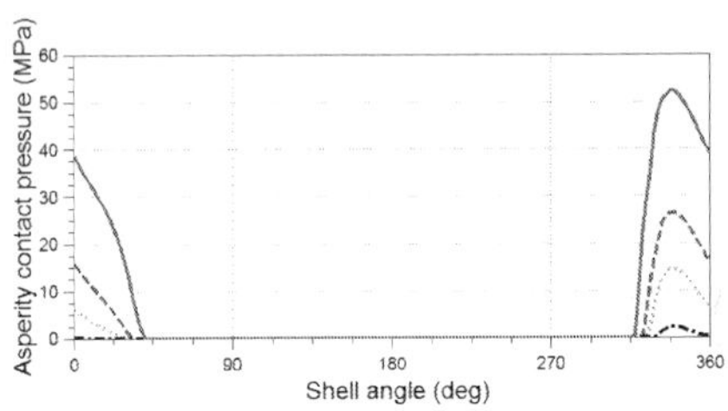

Figure 10. Comparison of the asperity contact pressures occurring at the outermost bearing shell edges for a load of 70 MPa for the different oils: SAE10 denoted as solid red line, SAE20 denoted as dashed blue line, SAE30 and SAE40 shown as green dotted line and as black dash-dotted line, respectively.

begins to reduce the benefit of the decreased hydrodynamic losses. As can be seen from the results, also for this case the prediction accuracy of the presented simulation method lies for all cases studied within the measurement uncertainty of the experimental data.

Finally, arriving at the lowest viscosity lubricant SAE10 this case represents the most severe operating conditions in terms of mixed lubrication that are investigated in this work. For the most severe case of the highest dynamic load of 70 MPa and the lowest journal speed studied of 2000rpm, already about 14% of the total power losses are caused by metal-metal contact. While the journal bearing might endure these conditions on the test-rig still for a rather long time, the amount of metal-metal contact is much to high to be allowed in real world applications. Significant wear occurs every load cycle which does not stabilize after a run-in period and leads, therefore, to a constant wear of the bearing shell. For the lower load case of 41 MPa, simulation still predicts no asperity contact and attributes all friction power losses to the hydrodynamic losses, see Fig. 9.

Finally, the ability of the presented method to predict the existence of metal-metal contact is put to test. For this task, a bearing durability test is investigated. For this test, the operating conditions are made even more severe by increasing the dynamic load to a maximum of 76 MPa at a journal speed of 3000rpm and increasing the oil inflow temperature of the SAE10 lubricant to 110°C. In comparison to the previous operating conditions with an oil inflow temperature of 80°C, this temperature increase causes the lubricant viscosity to decrease by more than 50% (see Fig. 6). These operating conditions lead consequently to bearing shell temperatures exceeding 130°C. As significant metal-metal contact occurs for these operating conditions, it can be detected by contact voltage measurements. For this measurement, a voltage is applied e.g. in form of a charged capacitor between the journal and the bearing. As the lubricant has only a poor electrical conductivity, the capacitor stays charged and the voltage remains unchanged. When metal-metal occurs, the capacitor can discharge due to the corresponding increased electrical conductivity; this process can be observed as change (decrease) of the voltage.

A comparison of the experimental contact voltage measurement and the predicted metal-metal contact is shown in Fig. 11 together with the applied dynamic load. It can be seen that when the load exceeds a certain threshold, metal-metal contact occurs. When compared to the results from simulation, the onset and the duration of the calculated metal metal contact agrees very well with the measured contact voltage data.

Overall, the presented simulation method appears to describe the actual processes in the journal bearing sufficiently well, as it predicts the friction moment accurately and reliably over a large range of working conditions, which range from purely hydrodynamic to significantly mixed lubrication.

Other important properties related to reliability in lubricated journal bearings are the peak oil film pressure (POFP) and the minimum oil film thickness (MOFT) [18, 20], that are depicted in Figs. 12 and 13 for the investigated lubricants.

As shown in Fig. 12, the POFPs change significantly from about 90 MPa to 120 MPa between the two different loads, but do not vary significantly between the different lubricants at the same load.

For a load of 41 MPa the results show that the MOFT is for all investigated oil-classes above 1.5 μm, which is the asperity contact threshold. Therefore, no metal-metal contact occurs, which can also be seen in the power losses shown in Fig. 9.

Further, it is interesting to note that the MOFT decreases by about 0.5 μm for every decrease in SAE-class; while the MOFT is considerably large with 3 μm at the point of maximum load for lubrication with SAE40, it decreases to about 2.5 μm and 2.0 μm for SAE30 and SAE20, respectively. For SAE10 the MOFT at the point of maximum load decreases further to about 1.5 μm; while from simulation still no asperity contact occurs for this case, in practical applications other effects not included here, like journal misalignment may lead to asperity contact.

The situation is quite different for a load of 70 MPa where all oils cannot avoid a certain amount of asperity contact and the MOFT consequently drops for all oils below 1.5 μm, however, for a different number of degrees crank angle. It is instructive to note that for a load of 70 MPa the MOFT changes not by the same amount between the different viscosity classes as for the 41 MPa load case. This can be explained by the choice of the elastic factor in

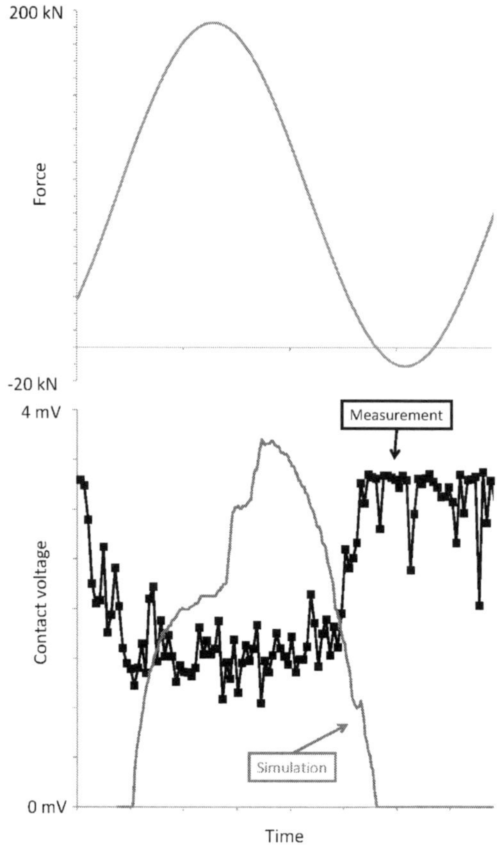

Figure 11. Plot of the measured contact voltage (black markers), whose decrease signals occuring metal-metal contact, in comparison to the metal-metal contact predicted from simulation (red line); as reference the blue curve depicts the applied load (top) for the bearing durability test.

the contact model that directly influences how much the asperity contact pressure increases for a given reduction in oil film thickness, see Eqs. (3) and (4). Due to the use of a large elastic factor, as discussed in Sec. 3.1.2, already small changes in oil film thickness lead to large changes in the asperity contact pressure and, therefore, increase significantly the total load carrying capacity.

In Fig. 10 this increase in asperity contact pressure is depicted over shell angle at the shell edge for lubrication with the different oils and a load of 70 MPa. As can be seen, the asperity contact pressure increases from a maximum of about 2.5 MPa for the SAE40-oil to a maximum

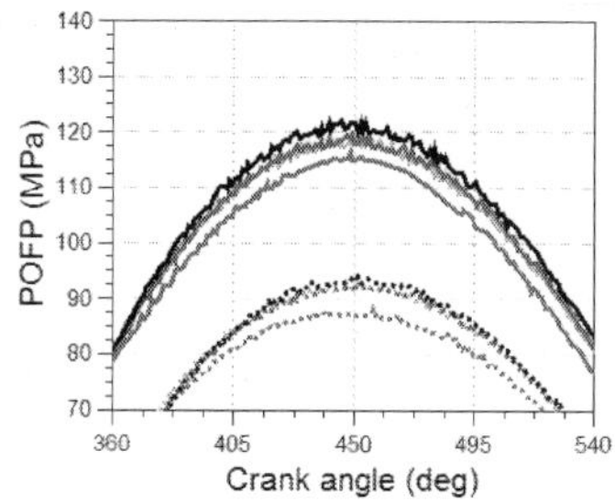

Figure 12. Plot of the calculated peak oil film pressures (POFP) occurring in the test bearing during the 90 degrees crank cycle when the load is applied (see Fig. 5); dashed lines denote the results for an applied load of 41 MPa, solid lines the results for an applied load of 70 MPa; the red curves represents the results for SAE10, the green for SAE20, the blue for SAE30 and the black curves show the results for lubrication with SAE40-oil.

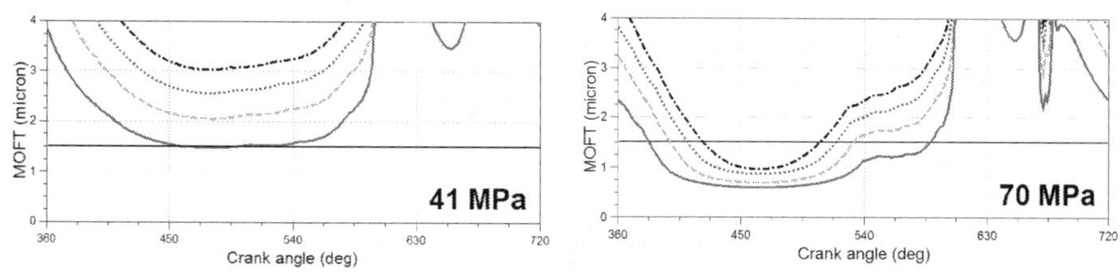

(a) (b)

Figure 13. Comparison of the minimum oil film thickness (denoted as MOFT, unit in μm) during a crank cycle in the test bearing for a load of (a) 41 MPa and (b) 70 MPa for 2000rpm journal speed calculated for the different oils: red solid line denotes SAE10, the green dashed line SAE20, the blue dotted line SAE30 and the black dash-dotted line SAE40. In addition the line at 1.5μm shows the threshold when asperity contact starts.

of 50 MPa for lubrication with the SAE10-oil and this explains the small MOFT changes that are seen for the different oils.

The presented asperity contact pressures occur only at the outermost bearing shell edges and drop sharply within a few millimetres in axial direction, as shown in Fig. 14 exemplary for lubrication with SAE10 and a load of 70 MPa. The concentration of asperity contact in these areas is a consequence of the elastic deformation of the shell and journal under load and, therefore, these areas are particularly subject to wear. However, the exact shape of these areas depends also on the stiffness of the supporting structures.

This result further demonstrates the necessity of using realistic surface profiles in the simulation as otherwise asperity contact in these areas, and consequently also the friction moment, is largely overestimated. Performing an actual simulation for a load of 70 MPa and a lubrication with SAE10 without the discussed surface profile for the test bearing, simulation predicts a power loss due to asperity contact of 389 W (averaged over a full crank cycle), which

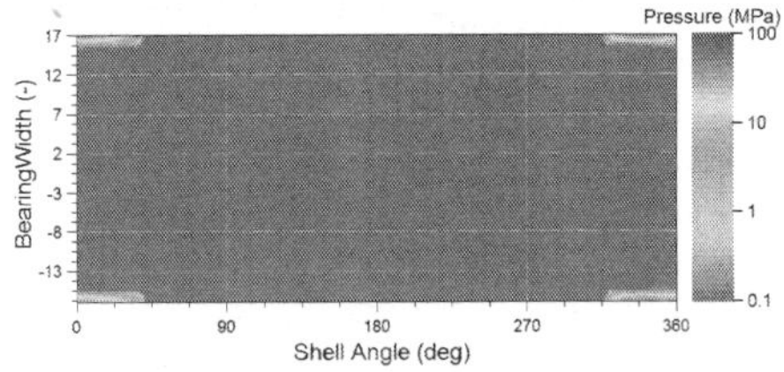

Figure 14. Map of the asperity contact pressures calculated for a load of 70 MPa and lubricated with SAE10-oil: the largest asperity contact pressures occur in very small areas at the bearing edges and drop sharply off within a few millimetres. The regions displayed in blue represent full film lubrication.

P_{ACP} [W]	P^{noWP}_{ACP} [W]	M_{Sim} [Nm]	M^{noWP}_{Sim} [Nm]	M_{LP06} [Nm]
64	389	4.7	6.9	(4.4-4.8)$\pm$ 0.5

Table 2. Summary of the average friction moments predicted by simulation for a load of 70 MPa and lubrication with SAE10: the simulation using a surface profile for the test bearing is denoted as M_{Sim} and denoted as M^{noWP}_{Sim} is the simulation without surface profile for the test bearing. The experimental values are denoted as M_{LP06} (range in brackets, $\pm$ measurement accuracy).

exceeds by more than a factor six the value obtained for the simulation with the wear profile, namely 64 W, as shown in Fig. 9.

Consequently, simulation predicts for a load of 70 MPa and lubrication with SAE10 an average friction moment of 6.9 Nm which exceeds the experimental values by about 50%, see Tab. 2.

3.1.6. Summary

Summarizing, for all investigated working conditions the presented simulation approach achieves very good agreement with the measured values and all calculated average friction moments lie within the experimentally found value range.

From a methodological point of view we find from the results that the shown accuracy can only be obtained reliably if the following important points discussed in this work are used in combination: a pressure dependent oil-model is of crucial importance as the dominant pressure thickening of the lubricant allows an easy and straightforward temperature estimation for the global oil temperature used in the EHD-calculation (this is shown in more detail also in [2]). Further, to avoid a large overestimation of the power loss due to metal-metal contact, deviations from the nominal perfect geometry of the bearing due to wear have to be taken into account and, further, a reduced surface roughness is efficient in combination with the Greenwood and Tripp model to take into account the conformal character of the bearing surfaces.

The origins of friction, asperity contact and hydrodynamic losses are discussed for the investigated lubrication cases and it is found that for the lower load, namely for 41 MPa, simulation predicts that no asperity contact occurs for lubrication with all oils studied.

Though the minimum oil film thickness found in simulation for SAE10 is very close to the threshold of asperity contact, all investigated oils are expected to provide sufficient lubrication for this load. The correspondingly calculated friction power losses demonstrate well the attainable friction reduction by choosing a lubricant with the optimum viscosity for a certain load; in particular, for a load of 41 MPa the friction power losses can be reduced from 518 W for SAE40 down to 340 W for SAE10, which is a reduction by 33%.

For the significantly more severe load case of 70 MPa, a certain amount of asperity contact is seen for all studied oils in the results from simulation. Indeed, the hydrodynamic friction power losses could be reduced from 626 W for the SAE40 oil to 453 W calculated for the SAE10 oil, which corresponds theoretically to almost a 30% reduction in friction. However, power losses caused by metal metal contact diminish this potential by a certain extent. Still, the total friction power losses are reduced from 626.5 W for SAE40 to 517 W for SAE10, which is still a reduction by almost 20%. This result demonstrates the efficiency in reducing friction by choosing a lubricant with reduced viscosity, so that, as long as measures to maintain reliability are taken (e.g. by using antiwear-additives in the lubricant), even a small amount of asperity contact might be tolerated to reduce the total friction power losses. This result is the starting point for the further work in Sec. 4

3.2. Including thermal processes - TEHD

Although the agreement of the previously presented simulation method with the experimental data is very much on spot and within the experimental accuracy for the whole range of working conditions studied, the neglect of local temperatures is a rough approximation of reality and its deviations to a model including local temperatures need to be quantified. In the following the isothermal simulation method is extended to a thermoelastohydrodynamic (TEHD) calculation to consider local temperatures. While critically important information like the bearing shell back temperatures was needed as input for the EHD simulation, these should now emerge from a representative thermal submodel. Due to the fact that the extension to TEHD is not straightforward and its thermal submodel adds a number of additional uncertain factors, the boundary conditions are chosen not only on a basis of physical arguments but also sufficiently distant from the oil film to minimize their influence on the results.

In the following the focus is on the extension of the isothermal simulation method to include local temperatures; therefore, all details regarding the experimental part are not repeated, but can be found in Section 3.1.

3.2.1. *Theory*

From a methodological point of view the TEHD-approach to journal bearings is well known; it adds the energy equation and heat equation to the set of differential equation that need to be solved together. While this approach represents a more complete picture of the physical processes in journal bearings, it has several severe drawbacks. While the dramatically increased cpu-time is more of a practical drawback that might be resolved by faster computers or faster solvers, the required thermal submodel needs a detailed knowledge of the thermal properties and of the heat flows of the test-rig. Therefore, subsection 3.2.4 is devoted to discuss its derivation.

Following [17] the Reynolds equation is combined with the energy equation to be able to take thermal processes into account. In a body shell fixed coordinate system such an extended Reynolds equation can be defined as [17, 23]

$$\begin{aligned} &-\frac{\partial}{\partial x}\left(\theta\alpha^2\frac{\partial p}{\partial x}\right)-\frac{\partial}{\partial z}\left(\theta\alpha^2\frac{\partial p}{\partial z}\right)+\\ &+\frac{\partial}{\partial x}\left(\theta\beta\right)+\frac{\partial}{\partial t}\left(\theta\gamma\right)=0, \end{aligned} \tag{5}$$

where x, z denote the circumferential and axial directions and θ the oil filling factor. α, β, γ are defined as

$$\begin{aligned} -\alpha^2 &= h^3\cdot\int_0^1\rho\left(\int_0^y\frac{y'}{\eta'}dy'-\frac{\int_0^1\frac{y'}{\eta'}dy'}{\int_0^1\frac{1}{\eta'}dy'}\cdot\int_0^y\frac{1}{\eta'}dy'\right)dy\\ \beta &= h\cdot U\int_0^1\rho\left(1-\frac{\int_0^y\frac{1}{\eta'}dy'}{\int_0^1\frac{1}{\eta'}dy'}\right)dy\\ \gamma &= h\int_0^1\rho\,dy, \end{aligned} \tag{6}$$

where the direction along the film height, y, is normalized to the oil film height h, thus the integration is carried out from 0 to 1. Further, the prime ($'$) indicates that the corresponding quantity depends on y'. U is the difference of the circumferential speeds of journal and bearing shell, $U = U_{\text{Shell}} - U_{\text{Journal}}$. η and ρ denote the oil viscosity and density, both are considered as pressure and temperature dependent.

The extended Reynolds equation (5) is solved together with the energy equation for the fluid film that is extended with the term $f(\text{Asp})$ to include heating due to asperity contact,

$$\begin{aligned} &\rho c_p\left\{\frac{\partial T}{\partial t}+u\frac{\partial T}{\partial x}+w\frac{\partial T}{\partial z}\right.\\ &\left.+\frac{1}{h}\left[v-y\left(\frac{\partial h}{\partial t}+u\frac{\partial h}{\partial x}+w\frac{\partial h}{\partial z}\right)\right]\frac{\partial T}{\partial y}\right\}\\ &+\frac{T}{\rho}\frac{\partial\rho}{\partial T}\left(\frac{\partial p}{\partial t}+u\frac{\partial p}{\partial x}+w\frac{\partial p}{\partial z}\right)-\frac{\kappa}{h^2}\frac{\partial^2 T}{\partial y^2}\\ &=\frac{\eta}{h^2}\left[\left(\frac{\partial u}{\partial y}\right)^2+\left(\frac{\partial w}{\partial y}\right)^2\right]+f(\text{Asp}), \end{aligned} \tag{7}$$

where κ denotes the thermal conductivity, u, v, w refer to the (x, y, z) components of the fluid velocity vector $\mathbf{v}$ at the corresponding fluid point and c_p represents the specific heat of the lubricant. Eq. (7) considers heat convection in all three dimensions, heat conduction in radial direction, compression and viscous heating. $f(\mathrm{Asp})$ denotes the heating due to asperity contact,

$$\begin{aligned} f(\mathrm{Asp}) &= \frac{\tau_{\mathrm{Asp}}}{h^3} U \\ \tau_{\mathrm{Asp}} &= \mu_{\mathrm{Bound}} p_{\mathrm{Asp}}, \end{aligned} \tag{8}$$

where μ_{Bound} is the boundary friction coefficient and p_{Asp} the asperity contact pressure calculated using the Greenwood and Tripp model with the same parameters as discussed in the previous section 3.1.

For the thermal analysis of the bearing shell and a part of its surrounding structure, the actual geometry is approximated by a cylinder; for this cylinder the heat equation

$$\rho_s c_s \frac{\partial T_s}{\partial t} - \kappa_s \Delta T_s = 0 \tag{9}$$

is solved with the continuity of the temperature and heat flow as boundary conditions. The subscript s indicates that the quantities apply to this cylinder. As outer boundary condition of this cylinder to its ambient

$$\frac{\partial T_s}{\partial n} + \frac{\lambda}{\kappa_s}(T_s - T_R) = 0 \tag{10}$$

is used, where T_R is a reference temperature and λ the heat transfer coefficient to the ambient.

Last, as second outer boundary condition of the oil film to the journal, the journal temperature T_J was assumed to be constant (isothermal boundary condition).

To calculate the oil temperature in the oil groove an energy balance equation is used

$$\rho c_p V_o \frac{\partial T_o}{\partial t} = \int_{S_o} \left(\kappa_s \frac{\partial T_s}{\partial r} - \kappa \frac{\partial T}{\partial y} \right) dS + \rho c_p \Phi (T_o' - T_o) \tag{11}$$

that takes into account that a certain amount of hot oil is carried over the cavitation zone and mixes with the cold oil supplied to the oil groove. Also heat conduction between oil and bearing shell is considered. V_o denotes the volume and S_o the area of the oil groove, T_o the actual oil temperature in the oil groove, T_o' the supply temperature of the oil and Φ the oil flow from the groove into the bearing.

3.2.2. *Simulation*

The experimental data for the basic lubricant properties, namely density and dynamic viscosity, were already shown in the previous Section 3.1.2. In addition also the thermal lubricant properties are needed for the TEHD-simulation: specific heat and heat conductivity. These properties have also been measured for the same lubricant samples in the OMV-laboratory and the measured data are shown for the used SAE20 and SAE40 lubricant in Fig 15.

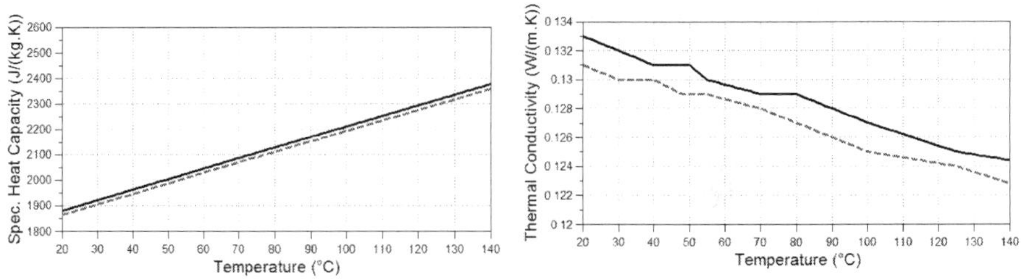

(a) (b)

Figure 15. Plot of (a) the specific heat c_p and (b) thermal conductivity κ of the used lubricant, shown as black solid line for SAE20 and as red dashed line for SAE40.

In the previous Section 3.1 the results from simulation were compared to the full range of values experimentally observed in repeated test runs; repeating the measurements with different, but brand new bearings resulted in measured total friction moments and temperatures that varied significantly between runs. Also, the quoted measurement uncertainty is significantly enlarged due to these repetitions.

For the following calculations the temperature data from a single test run for each lubricant/journal speed case is used to setup the thermal submodel of the TEHD calculation. Consequently, the results from the simulations are compared to these single test runs and the quoted smaller measurement uncertainty no longer contains the additional error due to repeating the test-runs with new bearings. Consequently, the measurement uncertainty is only about half of the previous measurement uncertainty, which represents a much stricter test for simulation.

3.2.3. *The thermal submodel*

On the test-rig a number of temperatures were measured together with the total friction moment that are either directly incorporated into the thermal submodel or used to validate it: notably the three bearing shell back temperatures, two con rod temperatures and the oil outflow temperature. The corresponding temperatures were measured by using thermocouple elements of type K that have an accuracy of $\pm 1°C$. The three bearing shell back temperatures were located at $\pm 45°$ circumferential angle from the vertical axis and the third in the middle at $0°$ circumferential angle as shown in Fig. 4 and represent, consequently, the temperatures in the high load area of the journal bearing.

On the con rod of the test-rig the temperature was measured in addition at 41mm distance from the bearing shell's inner diameter at two axial positions close to the high load zone. Last, the oil outflow temperature was also measured; Table 5 lists all these values.

A representative choice of the boundary conditions for the journal and to the ambient is of crucial importance as these affect the thermal flows in the system.

For the heat transfer coefficient λ of Eq. (10) a value of $120\frac{W}{m^2K}$ is used, which lies within the range of typically used values [7, 14, 28, 29]. The test con rod is made of 42CrMo4 and to describe its thermal behaviour a thermal conductivity of $\kappa_s = 46 \frac{W}{m\,K}$ and specific heat capacity of $c_s = 460.5 \frac{J}{kg\,K}$ are used (see also Table 3).

Further, the journal is assumed to be isothermal which sufficiently well approximates the real surface temperature as its circumferential speed is significantly larger than typical heat flow processes [17]. The journal temperature T_J is the only adjustable property in the simulation model and it is chosen such that the calculated bearing shell back temperatures averaged over one cycle, $\mathrm{T}_1^{\mathrm{Sim}}, \mathrm{T}_2^{\mathrm{Sim}}, \mathrm{T}_3^{\mathrm{Sim}}$, agree well with the corresponding experimental values ($\mathrm{T}_1^{\mathrm{Exp}}, \mathrm{T}_2^{\mathrm{Exp}}$, $\mathrm{T}_3^{\mathrm{Exp}}$), see Table 5.

Outward of the oil film 25 mm of the surrounding structure are taken into account in the thermal analysis; as reference temperature, denoted as T_R in Eq. (10), a value is used that was measured in the con rod of the test-rig at 41 mm distance from the bearing shell's inner diameter at two axial positions close to the high load zone. The used values for T_R, obtained by averaging the aforementioned two experimental values, are listed in Table 4.

The current model does not account for thermoelastic deformations and the consequent changes in bearing clearance. In the previous EHD-simulations a value for the bearing clearance was used that was measured experimentally at room temperature. While this simplification appears not to significantly affect the calculation of the total friction moment, the in this way calculated oil flow through the bearing does not agree well with the actual experimental oil flow. Therefore, the thermal power leaving the bearing with the oil-outflow cannot be expected to be calculated reliably. As consequence, it is attempted in the following to at least partially include thermoelastic effects by scaling the bearing clearance in the TEHD-calculations such that the oil flow through the bearing agrees with the experimentally controlled oil-flow of 2 litres per minute.

Further, it appears from the results that the such adapted bearing clearance has no other major impacts on the thermal submodel as the bearing shell temperatures experience only single degree changes.

As the amount of oil flow through the bearing is then properly included in the simulation, this further allows to calculate an approximative value[5] of the mean oil outflow temperature due to the power losses in the bearing and compare this value to the corresponding measured oil outflow temperature.

The mean temperature increase of the lubricant after flowing through the bearing, ΔT, is calculated using the thermal power leaving the bearing with the lubricant, $P_{\mathrm{oil\,out}}$, using

$$\Delta T = \frac{P_{\mathrm{oil\,out}}}{c_p \rho} \left(\frac{V}{t} \right)^{-1}, \tag{12}$$

where c and ρ denote the specific heat capacity and density of the lubricant, respectively, and $\frac{V}{t}$ is the oil volume flow through the bearing. The approximative mean oil outflow

[5] As we use a commercial software package we do not have access to the required numerical data to realize an exact scheme.

κ_S	$46 \frac{W}{m\ K}$
c_S	$460.5 \frac{J}{kg\ K}$
λ	$120 \frac{W}{m^2 K}$

Table 3. Table of the values used for the thermal conductivity κ, specific heat capacity c_s and the heat transfer coefficient λ of the test con-rod as denoted in Eqs. (9) and (10).

oil	SAE40		SAE20
speed [rpm]	2000	4500	2000
T_J [°C]	107	134	98
T_R [°C]	101	118	98

Table 4. List of the values for the reference temperature T_R, as employed in Eq. (10), and the constant journal temperature T_J used for the different investigated cases of journal speed and lubrication.

oil	SAE40		SAE20
speed [rpm]	**2000**	**4500**	**2000**
$T_{\text{oil supply}}$ [°C]	82.2	83.0	82.2
$T^{\text{Sim}}_{\text{oil groove}}$ [°C]	83.4	84.5	82.2
$T^{\text{Sim}}_{\text{oil out}}$ [°C]	94.0	109.7	89.6
$T^{\text{Exp}}_{\text{oil out}}$ [°C]	93±1	110±1	91±1
T^{Sim}_1 [°C]	101.6	123.3	95.6
T^{Exp}_1 [°C]	103±1	124±1	95±1
T^{Sim}_2 [°C]	107.6	136.8	98.5
T^{Exp}_2 [°C]	108±1	138±1	98±1
T^{Sim}_3 [°C]	106.9	137.3	98.7
T^{Exp}_3 [°C]	107±1	136±1	93±1

Table 5. Comparison of various measured temperatures with the corresponding calculated temperatures from simulation; T_1,T_2,T_3 denote the cycle averaged bearing shell back temperatures according to Fig. 4, where the superscript *Sim* denotes the calculated and *Exp* the measured values; $T^{\text{Sim}}_{\text{oil out}}$ and $T^{\text{Exp}}_{\text{oil out}}$ represent the calculated and measured oil outflow temperatures. Further, $T_{\text{oil supply}}$ denotes the oil supply temperature (same for simulation and test-rig) and $T^{\text{Sim}}_{\text{oil groove}}$ denotes the calculated oil temperature in the oil groove (see text). Finally, *oil* denotes the lubricant and *speed* the journal speed.

temperature is then calculated by

$$T_{\text{oil out}} = T_{\text{oil supply}} + \Delta T. \tag{13}$$

As the lubricant properties c_p and κ themselves depend on the temperature, this procedure is only approximative.

In the following a number of results for different working conditions ranging from full film lubrication (SAE40 and 4500rpm journal speed) to the onset of mixed lubrication (SAE20 and 2000rpm journal speed) are discussed.

3.2.4. Results for the thermal submodel

As shortly outlined in the introduction of this section, extending the EHD-simulation to TEHD brings significant additional complexity due to the thermal submodel that has to be

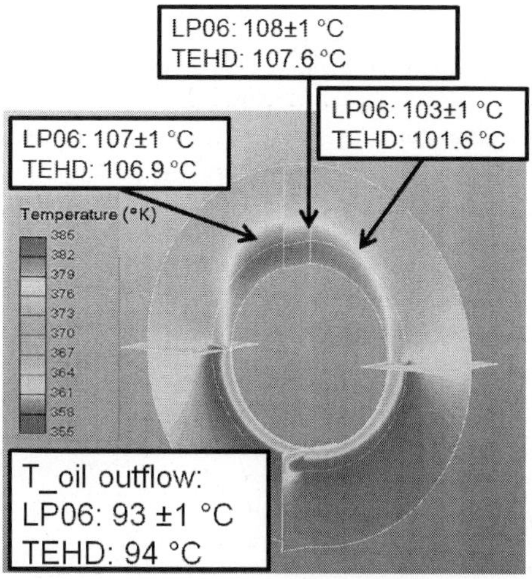

Figure 16. Plot of the temperatures occuring in the oil film and surrounding bearing structure in the axial center of the bearing for the exemplary case of lubrication with SAE40 and a load of 70 MPa at a journal speed of 2000rpm. *LP06* denotes the values measured on the test-rig and *TEHD* the values calculated with the TEHD-method. In addition the mean oil outflow temperature calculated from TEHD, $T_{\text{oil outflow}}$, is compared to the experimentally measured value. The thin white line represents the border between oil film and bearing structure (the oil film is drawn disproportionately large).

employed. As the local temperatures directly influence the lubrication properties within the bearing the thermal submodel needs to be validated with experimental measurements. Table 5 contains for all cases studied the temperatures calculated at several points of the test-rig and allows to compare these to the corresponding measured temperatures, in particular Fig. 16 shows a graphical comparison for lubrication with the SAE40 lubricant and a journal speed of 2000rpm. As previously mentioned it is the aim to choose the journal temperature for each case such that the calculated bearing shell temperatures agree within about 1°C with the measured temperatures; this could be achieved for all cases except for the SAE20 result, where one of the three bearing shell temperatures deviates from the measured one by about 5°C.

As the temperature of the whole test-rig lies above the oil inflow temperature only the friction power losses in the bearings heat the system. Therefore, as an additional validation the calculated (see Eq. 12 and 13) and measured oil outflow temperatures are compared in Table 5. As one can see from these results the predicted mean oil outflow temperature agrees for all cases within 1°C with the measured mean oil outflow temperatures. Summarizing, the results confirm that the thermal submodel is suitably chosen and representative for the test-rig.

For sake of completeness Table 5 also lists the actual oil temperatures in the oil groove that are calculated for each case as defined by Eq. (11). From the results one finds that the actual oil temperature in the oil groove is only weakly affected by the surrounding hotter structure as the temperature differences are very small for all cases.

3.2.5. Results for the friction power losses

In the following the results obtained with the various methods are discussed in terms of friction prediction; to this task the results from TEHD are compared to the results obtained from EHD. As the TEHD comes with a largely increased cpu-time requirements, only the test-bearing is investigated using TEHD in the following.

As briefly mentioned in Sec. 3.1, the discussed EHD-simulation model employs the same lubricant temperature for test and support bearings. This restriction needs to be removed first to allow consequently a detailed comparison of the results for the test-bearing obtained from different simulation methods.

In [1] a new method is proposed that allows to easily calculate a suitable oil film temperature for the isothermal EHD-calculation. The procedure combines the measured bearing shell temperature, T^{Exp}, and the supplied oil temperature, $T_{\mathrm{oil\ supply}}$, to

$$T^{\mathrm{comp}} = T^{\mathrm{Exp}} - \frac{T^{\mathrm{Exp}} - T_{\mathrm{oil\ supply}}}{4} \tag{14}$$

and yields the same accurate results as can be seen in Figs. 17 and 18, but has the advantage of being applicable to single bearings. For reference also shown are the results that are obtained from the same EHD model but with the previously in Sec. 3.1 discussed temperature estimation for all three combined bearings.

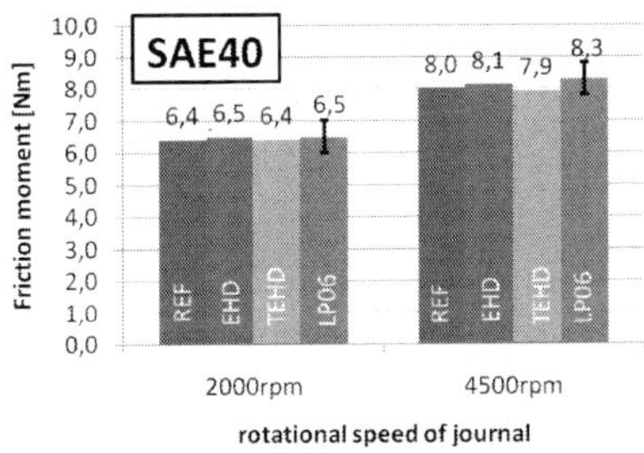

Figure 17. SAE40, 70 MPa: comparison of the total friction moments calculated by EHD and TEHD to the experimental values measured on the test-rig LP06; the black bars denote the measurement accuracy of ±0.5 Nm. Denoted as *REF*, the values calculated by the previous temperature estimation method of Sec. 3.1 are shown for reference (see text).

As a first test for both new methods (TEHD-simulation of the test-bearing and the new method to obtain a representative lubricant temperature for the EHD-simulation of the support bearings), again the total friction moment for all three bearings is compared to the experimental data.

Starting with working conditions within full film lubrication, Fig 17 shows the total friction moment calculated by EHD and TEHD and gives a comparison to the experimental values for SAE40 and two different journal speeds. As can be seen the TEHD-results do not show systematic deviations that might be expected due to the inclusion of local temperatures. In fact, the results are almost identical; for a journal speed of 2000rpm 6.4 Nm are calculated by TEHD in comparison to the 6.5 Nm predicted by EHD and the 6.5±0.4 Nm measured on the test-rig. For an increased journal speed of 4500rpm the results are again very close with 7.9 Nm and 8.1 Nm predicted by TEHD and EHD, respectively, compared to the 8.3±0.5 Nm measured experimentally.

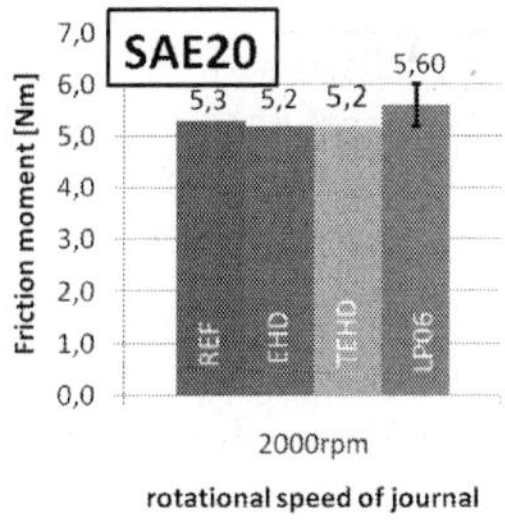

Figure 18. SAE20, 70 MPa, 2000rpm: comparison of the total friction moments calculated by EHD and TEHD to the experimental value measured on the test-rig LP06; the black bars denote the measurement accuracy of ±0.5 Nm. Denoted as *REF*, the values calculated by the previous temperature estimation method of Sec. 3.1 are shown for reference (see text).

For the case of SAE20 and a journal speed of 2000rpm weak mixed lubrication occurs. As Fig. 18 depicts, even the same total friction moment is calculated by both methods for this working condition.

For reference Figs. 17,18 also show the results calculated with the previously in Sec. 3.1 discussed method, where the same lubricant temperature was used in the EHD-simulations of all three bearings. This procedure is not suited to discuss properties of the individual bearings and it is only shown for completeness. As can be seen from the results for the total friction moment, the previously proposed method is very accurate.

While the total friction moment arises from all three bearings, the details of the calculated friction power losses are discussed in the following focusing on the test-bearing only. As was found in Sec. 3.1, for a load of 70 MPa and a journal speed of 4500rpm SAE40 oil provides sufficient lubrication such that the test bearing stays in the full film lubrication regime. This is also supported by the results obtained from the TEHD-calculation, where the individual contributions to the power losses are depicted in Fig. 19.

For the case of lubrication with SAE40 and a journal speed of 2000rpm the hydrodynamic power losses calculated by both methods are very close with a difference of 5%. The TEHD-model takes into account the hotter bearing shell temperatures at the edges of the bearing in the high load zone and, as consequence, a vanishing amount of metal-metal contact occurs. For the case of a journal speed of 4500rpm the differences between the results from EHD and TEHD increase to about 9%.

For the case of weak mixed lubrication (lubrication with SAE20 and a journal speed of 2000rpm) the results (shown in Fig. 20) are again very close and differ only by about 5%

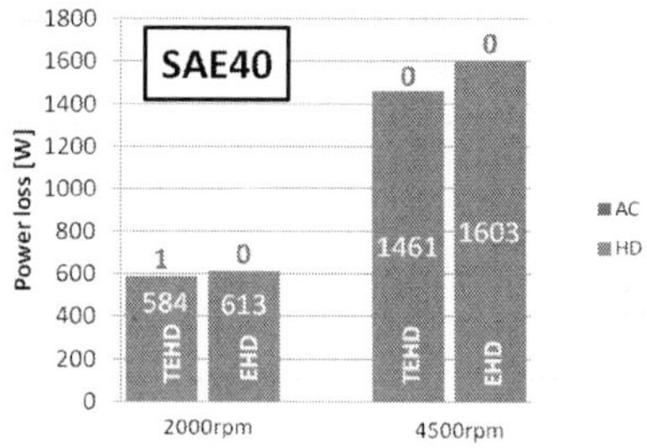

Figure 19. Comparison of the power losses in the test-bearing calculated with EHD and TEHD, respectively, for lubrication with SAE40 and a journal speed of 2000rpm (left) and 4500rpm (right). *HD* denotes hydrodynamic power losses and *AC* power losses caused by metal-metal contact. The numbers denote the quantities of the corresponding contributions.

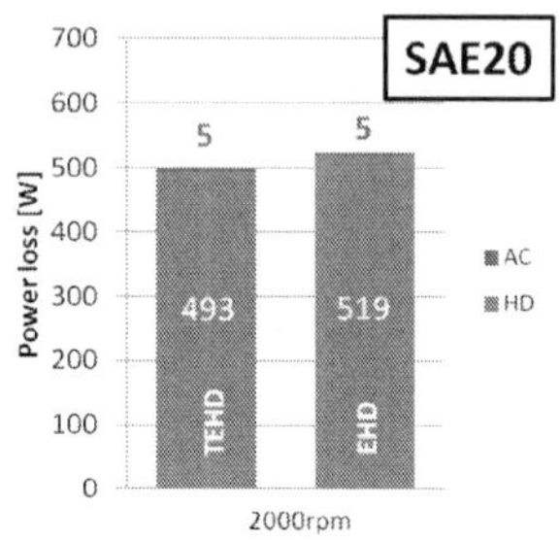

Figure 20. Comparison of the power losses in the test-bearing calculated with EHD and TEHD, respectively, for lubrication with SAE20 and a journal speed of 2000rpm. *HD* denotes hydrodynamic power losses and *AC* power losses caused by metal-metal contact. The numbers denote the quantities of the corresponding contributions.

in terms of the hydrodynamic power loss. It is interesting to note that both methods predict about the same (small) amount of power loss due to metal-metal contact.

3.2.6. *A detailed TEHD vs. EHD comparison*

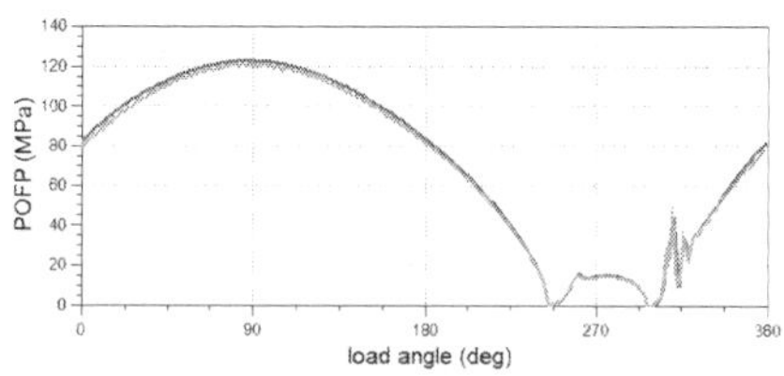

Figure 21. Exemplary plot of the peak oil film pressures (POFP) calculated with TEHD (red line) and with EHD (green line) for lubrication with SAE40 and a load of 70 MPa at a journal speed of 2000rpm.

However, while so far the differences between the EHD and TEHD-results have been rather small, the calculated lubricant viscosities are expected to be notably different. Although the calculated hydrodynamic pressures in the journal bearing match closely as Fig. 21 shows, differences arise due to different local temperatures.

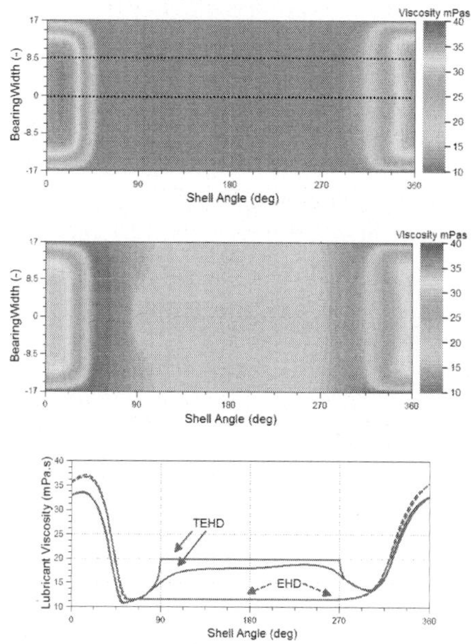

Figure 22. Plot of the oil viscosity in the test bearing at the point of maximum load for lubrication with SAE40 and a journal speed of 2000rpm, calculated with EHD (top) and TEHD (middle) using the same scale for comparison. For a detailed comparison the plot at the bottom shows the lubricant viscosity from EHD (dashed lines) and from TEHD (solid lines) over shell angle across 8.5mm (blue lines) and 0mm (red lines) bearing width (position of these cuts shown as black dashed lines in the plot on top). The rectangle shaped area with cool (thick) lubricant in the TEHD-results is the oil supply groove that cannot be taken into account directly in the EHD-calculation.

As can be seen in Fig. 22, the lubricant viscosities show significant differences. For the shown two different axial cuts through the high load zone the EHD-result predicts a maximum lubricant viscosity that is about 9% higher than the corresponding value calculated by TEHD. This 9% difference in lubricant viscosity corresponds roughly to a difference of only about 4°C in lubricant temperature (for SAE40) (see lubricant data in [3]). Consequently, this further confirms that the EHD-method with the proposed temperature compensation describes well the lubrication conditions also in the high load zone of the bearing.

3.2.7. Dynamic behaviour of the local oil film temperatures

Further, to quantify the approximation of the true bearing by using an isothermal EHD-model the lubricant temperatures close to the bearing shell are plotted in Fig. 23 for the exemplary

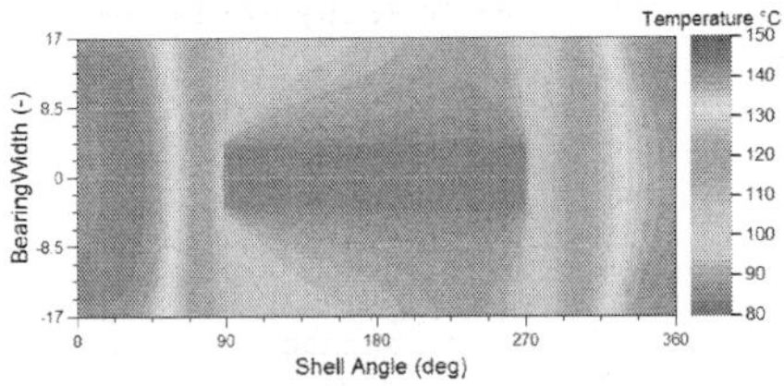

Figure 23. Plot of the calculated oil temperatures adjacent to the bearing shell for lubrication with SAE40 and a journal speed of 4500rpm at the point of maximum load (see Fig. 5); the actual maximum oil temperature in this plot is 144°C (see also Fig. 24).

case of SAE40 and a journal speed of 4500rpm, at the point of maximum load. These results show that spatially the oil temperature is almost constant in the high load area of the bearing (small temperature gradients in large areas of the bearing). Of course as the load changes the power losses in the oil film change and, consequently, local heating changes. Therefore, the dynamic change of the maximum lubricant temperature over a load cycle is shown in Fig. 24 for all cases studied. From these results one finds that the temperature variations in the journal bearing during one load cycle stay with about 2°C and are, therefore, also rather small. The observation that the temperature changes in the journal bearing are very small is in fact not new, it was e.g. discussed in [17]. The reasons behind this behaviour is on one hand that most of the heat (about 70% for the cases studied here) leaves the system with the oil and on the other hand that the thermal conductivity of steel is rather poor.

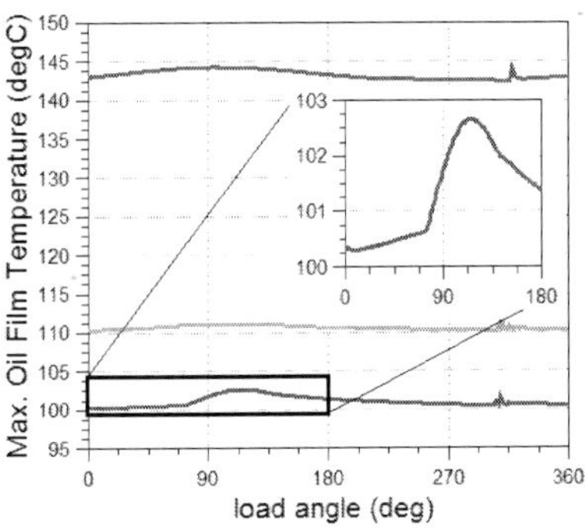

Figure 24. Plot of the maximum oil temperatures evolving over a load cycle for lubrication with SAE40 and a journal speed of 4500rpm, shown as red curve, for 2000rpm shown as green curve, and for lubrication with SAE20 and a journal speed of 2000rpm shown as blue curve. As metal-metal contact occurs in the SAE20-case this result consequently shows also the largest changes in the maximum oil film temperature. However, as the inset shows also for this case the maximum temperature change during one load cycle stays within about 2°C.

As was argued already in Sec. 3.1, the high load area dominates the power losses in the bearing due to the occurrence of high hydrodynamic pressures of more than 100 MPa in this area.

oil	run	speed [rpm]	M_{EHD} [Nm]	M_{TEHD} [Nm]	M_{LP06} [Nm]
SAE40	364	2000	6.4	6.6	6.5 ±0.4
		4500	8.0	8.0	8.3 ±0.5
SAE20	377	2000	5.3	5.3	5.6 ±0.5

Table 6. Table of the total friction moments predicted by EHD, denoted as M_{EHD}, by TEHD, denoted as M_{TEHD}, and the corresponding total friction moment (± measurement accuracy), denoted as M_{LP06}, measured for the indicated run of which the temperature data was used to setup the thermal submodel of the TEHD-calculation. Speed denotes the journal speed and all data apply to the transient load with a maximum value of 70 MPa.

These pressures lead to a more than threefold viscosity increase of the lubricant in this part of the bearing. As the lubricant temperature varies only weakly in this area (both spatially as well as dynamically), it is indeed well described by the previously proposed isothermal EHD-approach [3] that takes into account the pressure dependence of the lubricant viscosity.

3.2.8. Summary

Within this section the previously discussed EHD-based simulation model was extended to TEHD to investigate differences arising due to the neglect of local temperatures. With the experimentally obtained temperature data for each working condition it was possible to validate the thermal submodel of the TEHD-method. By choosing appropriate thermal boundary conditions it was shown that the TEHD-method is able to predict the occurring temperatures at several different points of the test-rig, namely three bearing shell back temperatures and the oil outflow temperature, very accurately within 1°C of the measured temperatures (with only one exception where the temperature difference is still rather small with 5°C).

Concerning the ability to predict friction power losses in journal bearings, the results indicate that the considerably simpler EHD-approach appears to be sufficient to reliably and accurately predict these losses for full film lubrication. Also for weak mixed lubrication, as was studied using the SAE20 lubricant, the neglect of local temperatures results again only in small deviations that do not significantly affect the studied properties.

As was shown from the TEHD-results, the temperature in the high load zone stays almost constant throughout the load cycle and is also almost constant spatially within this high load zone; due to this temperature stability the EHD-method works very well.

In conclusion, the results demonstrate that even for highly loaded journal bearings as they are common in today's diesel and gasoline engines the EHD-method is still well suited to study the bearing working conditions, both in terms of the generated friction power losses and to investigate the occurrence of metal-metal contact. The inclusion of local temperatures in the form of an extension to TEHD does not give significant changes for full film and weakly mixed lubrication for the properties studied. It is, however, expected to be of more importance in mixed lubrication with significantly more metal-metal contact. However, such severe operating conditions do not occur in serially produced engines and are, therefore, not within the scope of this work.

4. Friction reduction for engines - a practical example

In the following, the potential for friction reduction in the journal bearings of the crank train shall be analysed for a modern four cylinder passenger car turbodiesel engine lubricated with common multi-grade oils using the isothermal method discussed in Sec. 3.1. In particular, Styrene-Isoprene-Copolymer (SICP)-additive enhanced oils are considered in the following (in terms of shear rate dependency of the lubricant). For all variants, the friction will be calculated as sum of all five crankshaft main bearings and four big end bearings at full-load operation with a peak cylinder pressure of 190 bar, which leads to specific bearing loads of up to about 50 MPa for the main bearings and to about 90 MPa for the big end bearings. Further, the dynamic oil supply for the big end bearings is realistically represented in the simulation as oil supply network.

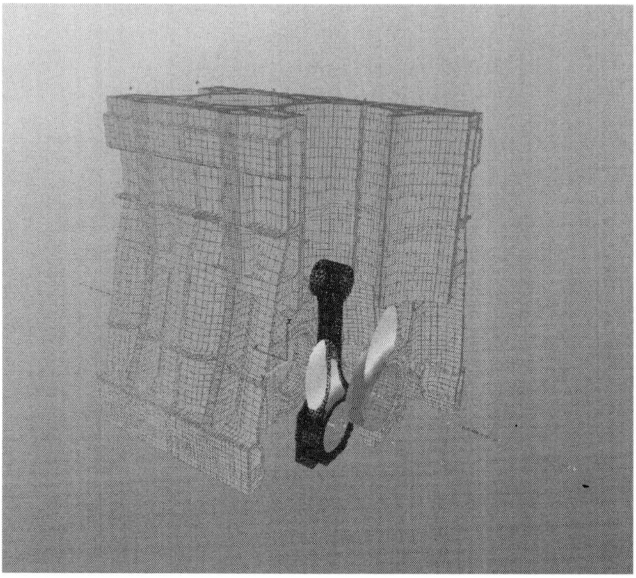

Figure 25. Plot of a part of the inline four cylinder engine for which the calculations are carried out; it shows the locations of the main and big end bearings which are shown with examplary oil film pressure distributions shown as 3D-plot.

4.1. Finding a friction optimized solution

In the following basic example, easily modifiable parameters such as bearing shell width and viscosity grade (SAE-class) of the engine oil are in the focus. The savings potential derived from the reduction of the bearing shell width is set with the reduction of the oil-filled volume and the use of low viscosity oils directly influences the viscosity losses. Both measures reduce the load capacity of the bearings. Therefore, it is crucial to identify occurring mixed lubrication in order to find a low friction solution which does not impair the bearing lifetime through emerging mixed lubrication. To illustrate the influence of the bearing shell width on the friction losses, widths of 21mm and 16mm are studied in addition to the original width of 18mm. Further, the original lubricant-grade 10W40 is reduced to 0W30 and for a specific case to 0W20. Two different oil temperatures are considered in the simulation because in everyday life not only warm engine conditions exist, but also low oil temperatures occur specifically with low exterior temperatures or for short driving distances. For the warm engine operation 100°C and for the cool operation 40°C oil temperature is considered.

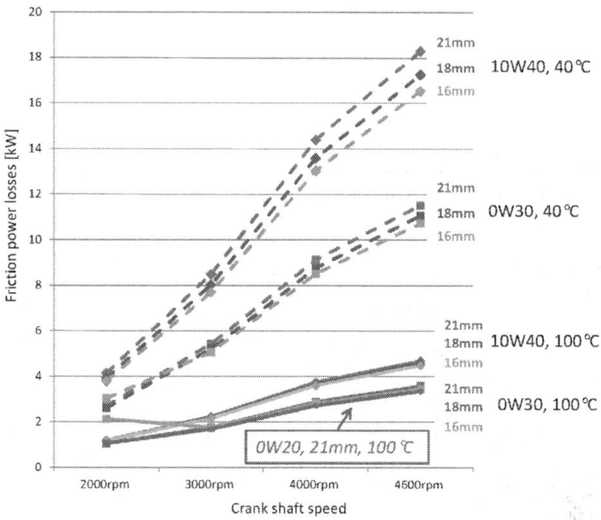

Figure 26. Friction power losses in the journal bearings at various speeds for different bearing shell widths, lubricant viscosity grades and operating temperatures.

Fig. 26 shows the summary of all results evaluated through the total friction power losses for all journal bearings at different speeds. Most notable from the results is that the oil temperature impacts the friction performance most intensely which is taken advantage of by many modern engines through a reduction of the oil volume in the oil sump and the related quicker warming-up of the engine. In particular, the friction power losses are cut in half throughout the entire speed range by increasing the lubricant temperature from 40°C to 100°C. Furthermore, it is shown that the different oil viscosities have a much stronger effect than the changes in bearing width. Referring to the example a savings potential is found to be in average 35% for the cold and still 20% for the hot case through changing from 10W40 to 0W30. In contrast, using more narrow bearings leads to a reduction of the losses by 9% for the cold case and a high speeds, but yields only a reduction of max 3% of the total journal bearing losses for the studied hot lubricant temperature. This minimal impact can be employed to use even lower viscosity oil for the engine and - while the load carrying capacity by utilizing wider bearings needs to be restored for this lubricant - a net reduction of friction power losses can be achieved. Fig. 26 also shows how the presented method assists in identifying potential issues of mixed lubrication: for the case of a 16mm wide bearing and lubrication with 0W30 there occurs for a lubricant temperature of 100°C already significant metal-metal contact at 2000rpm which leads to a significant rise in friction for this engine and potentially to problems in the operating reliability. However, with an enlarged bearing width even lower viscosity oil can be used; for the case presented the optimum is a low viscosity 0W20 oil combined with a broader bearing shell, in this case 21mm. Thereby, in comparison to the original configuration with 18mm bearings and 10W40 oil, the journal bearing losses can be reduced by 10% at 2000rpm and by approximately 30% at 4600rpm despite the significantly wider bearing shells.

4.2. Conclusion

The results show that small changes in the bearing geometry bear no significant impact on the friction losses in the journal bearings. However, the use of a low viscosity lubricant holds obvious advantages in regards to a reduction of these losses, despite the need of wider bearings to retain the bearing load capacity. In the presented example this combination of low viscosity lubricants with wider bearings revealed itself as optimal and proves approximately 10-30% decreased losses in comparison to the initial situation. Alternatively, if more complex in design, the increase in size of the journal bearing diameter and the therefore necessary larger journal diameter brings advantages also in regards to the NVH performance due to the increased stiffness of the crankshaft. Further measures for friction reduction like an on-demand oil supply could potentially also attain significant savings and be analysed through the presented model.

While this basic example of friction reduction in engines displays the efficiency of various measures, it is important to emphasise that the choice of the optimum lubricant affects the whole engine and the other major source of mechanical losses, namely the piston assembly, challenges with (partly) opposing requirements to the lubricant. In this sense, the optimum choice of the lubricant in terms of friction reduction shall only be taken under consideration of the complete system.

Acknowledgment

The authors would like to acknowledge several very interesting discussions on friction related topics and want to express their gratitude in particular to C. Forstner (MIBA Bearing Group), F. Novotny-Farkas (OMV Refining & Marketing GmbH), A. Skiadas (K & S Gleitlager GmbH) and O. Knaus (AVL List GmbH).

Further, the authors acknowledge the kind permission of the MIBA Bearing Group and the OMV Refining & Marketing GmbH to publish the results and the financial support of the 'COMET K2 - Competence Centers for Excellent Technologies Program' of the Austrian Federal Ministry for Transport, Innovation and Technology (BMVIT), the Austrian Federal Ministry of Economy, Family and Youth (BMWFJ), the Austrian Research Promotion Agency (FFG), the Province of Styria and the Styrian Business Promotion Agency (SFG).

Author details

H. Allmaier, C. Priestner, D.E. Sander and F.M. Reich
Virtual Vehicle Competence Center, Austria

5. References

[1] Allmaier, H., Priestner, C., Reich, F., Priebsch, H., Forstner, C. & Novotny-Farkas, F. [2012a]. Predicting friction reliably and accurately in journal bearings - extending the simulation model to TEHD, *Tribology International* 58 (2013): 20-28.

[2] Allmaier, H., Priestner, C., Reich, F., Priebsch, H., Forstner, C. & Novotny-Farkas, F. [2012b]. Predicting friction reliably and accurately in journal bearings - the importance of extensive oil-models, *Tribology International* 48: 93–101.

[3] Allmaier, H., Priestner, C., Six, C., Priebsch, H., Forstner, C. & Novotny-Farkas, F. [2011]. Predicting friction reliably and accurately in journal bearings–a systematic validation of simulation results with experimental measurements, *Tribology International* 44(10): 1151–1160.

[4] Bair, S., Jarzynski, J. & Winer, W. [2001]. The temperature, pressure and time dependence of lubricant viscosity, *Tribology International* 34(7): 461–468.

[5] Bukovnik, S., Offner, G., Caika, V., Priebsch, H. & Bartz, W. [2007]. Thermo-elasto-hydrodynamic lubrication model for journal bearing including shear rate-dependent viscosity, *Lubrication Science* 19(4): 231–245.

[6] European Commission [2010]. Reducing CO_2 emissions from passenger cars. URL: *http://ec.europa.eu/clima/policies/transport/vehicles/cars/index_en.htm*

[7] Fatu, A., Hajjam, M. & Bonneau, D. [2006]. A new model of thermoelastohydrodynamic lubrication in dynamically loaded journal bearings, *Journal of tribology* 128: 85.

[8] Fontaras, G. & Samaras, Z. [2010]. On the way to 130g CO_2/km–estimating the future characteristics of the average european passenger car, *Energy Policy* 38(4): 1826–1833.

[9] Greenwood, J. & Williamson, J. [1966]. Contact of nominally flat surfaces, *Proceedings of the Royal Society of London. Series A, Mathematical and Physical Sciences* 295: 300–319.

[10] Hall, D. & Moreland, J. [2001]. Fundamentals of rolling resistance, *Rubber chemistry and technology* 74: 525.

[11] Holmberg, K., Andersson, P. & Erdemir, A. [2012]. Global energy consumption due to friction in passenger cars, *Tribology International* 47: 221–234.

[12] International Energy Agency [2006]. World energy outlook.

[13] International Energy Agency [2008]. Worldwide trends in energy use and efficiency key insights from IEA indicator analysis.

[14] Kim, B. & Kim, K. [2001]. Thermo-elastohydrodynamic analysis of connecting rod bearing in internal combustion engine, *Journal of tribology* 123: 444.

[15] King, J. [2008]. The King review of low-carbon cars part I: the potential for reducing CO_2 emissions from road transport, *ISBN* 978: 1–84532.

[16] Körfer, T., Kolbeck, A., Schnorbus, T., Busch, H., Kinoo, B., Henning, L. & Severin, C. [2011]. Fuel consumption potential of the passenger car diesel engine after EURO 6, 32. *Internationales Wiener Motorensymposium*, VDI, Vienna, Austria, pp. 99–122.

[17] Krasser, J. [1996]. *Thermoelastohydrodynamische Analyse dynamisch belasteter Radialgleitlager (Thermoelastohydrodynamic analysis of dynamically loaded journal bearings)*, PhD thesis, Graz University of Technology.

[18] Mittwollen, N. & Glienicke, J. [1990]. Operating conditions of multi-lobe journal bearings under high thermal loads, *Journal of Tribology* 112: 330.

[19] Mufti, R. & Priest, M. [2009]. Technique of simultaneous synchronized evaluation of the tribological components of an engine under realistic conditions, *Proceedings*

of the Institution of Mechanical Engineers, Part D: Journal of Automobile Engineering 223(10): 1311–1325.

[20] Okamoto, Y., Kitahara, K., Ushijima, K., Aoyama, S., Xu, H. & Jones, G. [2000]. A study for wear and fatigue on engine bearings by using EHL analysis, *JSAE Review* 21(2): 189–196.

[21] Patir, N. & Cheng, H. [1978]. An average flow model for determining effects of three-dimensional roughness on partial hydrodynamic lubrication, *ASME, Transactions, Journal of Lubrication Technology* 100: 12–17.

[22] Patir, N. & Cheng, H. [1979]. Application of average flow model to lubrication between rough sliding surfaces, *ASME, Transactions, Journal of Lubrication Technology* 101: 220–230.

[23] Priebsch, H. & Krasser, J. [1997]. Simulation of the oil film behaviour in elastic engine bearings considering pressure and temperature dependent oil viscosity, *Tribology Series* 32: 651–659.

[24] Priestner, C., Allmaier, H., Priebsch, H. & Forstner, C. [2012]. Refined simulation of friction power loss in crank shaft slider bearings considering wear in the mixed lubrication regime, *Tribology International* 46(1): 200–207.

[25] Priestner, C., Allmaier, H., Reich, F., Forstner, C. & Novotny-Farkas, F. [2012]. Friction in highly loaded journal bearings, *MTZ* 4: 310–315.

[26] Smith, R. [2008]. Enabling technologies for demand management: Transport, *Energy policy* 36(12): 4444–4448.

[27] Tian, T. [2002]. Dynamic behaviours of piston rings and their practical impact. part 1: ring flutter and ring collapse and their effects on gas flow and oil transport, *Proceedings of the Institution of Mechanical Engineers, Part J: Journal of Engineering Tribology* 216(4): 209.

[28] Wang, Y., Zhang, C., Wang, Q. & Lin, C. [2002]. A mixed-TEHD analysis and experiment of journal bearings under severe operating conditions, *Tribology international* 35(6): 395–407.

[29] Zhang, C. [2002]. TEHD behavior of non-newtonian dynamically loaded journal bearings in mixed lubrication for direct problem, *Journal of Tribology* 124: 178.

Chapter 10

Theories on Rock Cutting, Grinding and Polishing Mechanisms

Irfan Celal Engin

Additional information is available at the end of the chapter

http://dx.doi.org/10.5772/56046

1. Introduction

Tribological research studies including cutting, abrading and polishing mechanisms have firstly started with metals, metal cutting theories and formulas have been developed and then applications on rock material have started. In this part of the book, natural stone cutting, abrasion and polishing mechanisms are compiled and presented as a summary.

Processes of cutting, grinding and polishing natural stones are made as a result of grinding-abrading mechanism developed on the use of different abrasive grains (mostly diamond and SiC). Wear intensity is named as cutting, abrading or polishing according to the speed, chip size and situation of obtained surfaces.

No matter which cutting machine is used, generally cutting process of natural stones are done with the use of segments that are obtained through sintering of diamond grains and metal powders. Industrial diamond grains in these segments rubbed against the material to be cut with a certain force and material is removed, and as a result, the material is cut along this surface as the material is removed as much as segment width.

In the stage of abrading and polishing of natural stones, products called grinding stone containing SiC grains are generally used. Intenseness of material removal from natural stone surfaces can be arranged by changing grain size of this abrasive and magnitude of pressure intensity. When relatively coarser grains and higher pressures are chosen, coarse abrading process is obtained while slight abrading and polishing is obtained when slighter grains and lower pressures are chosen.

2. The basic wear mechanisms emerging all types of materials

Wear is described in the literature as the loss of material as a result of the change in the shape of friction surfaces. Many researchers have stated that there are 4 main wear mechanisms causing the loss of material. These are adhesive wear, abrasive wear, and corrosive wear and wear resulting from surface fatigue (Archard, 1953; Moore, 1975; Suh and Saka, 1978; Williams, 1994; Summer, 1994). Similarly, many researchers have classified wear as heavy wear and light wear according to the wear magnitude.

A basic equation about wear is developed by Archard (1953). According to Archard (1953), wear on friction surfaces (w), is directly proportional to applied load (W) while inversely proportional to the strength of material (H).

$$w = K \, x \, \frac{W}{H} \tag{1}$$

K which is non-dimensional in here is expressed as wear coefficient. This coefficient is changed into k=K/H including strength; this is the dimensional wear coefficient which is more widely accepted in engineering. This coefficient represents the volumetric wear (mm^3) resulting from the shift in unit distance (m) under unit load (N).

When two materials are rubbed against each other, stresses on touch point can easily reach yield point. With the shearing effect of lateral force, material transfers from the surface of soft material to the surface of hard piece and sticks on. Wear developed this way is called adhesive wear. A simple demonstration of this is presented by Archard (1953) (Figure 1):

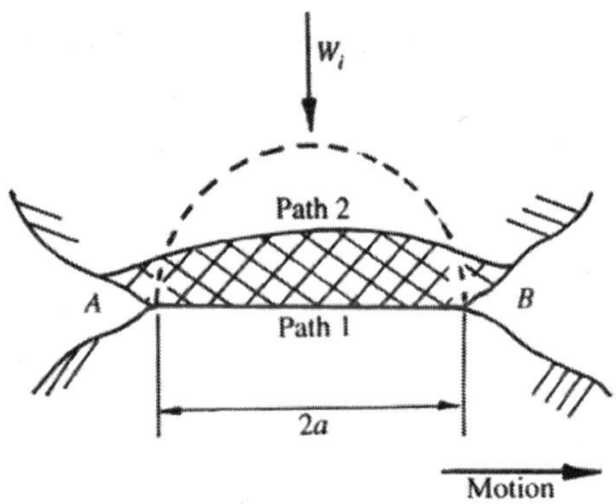

Figure 1. Material wear caused by the adhesion on friction surfaces (Archard, 1953)

Here, the diameter of contact point is shown as 2a, applied load is shown as W. It is thought that moving will be along the way shown as Path 2. For convenience, the part that will be abraded is assumed to be in the shape of a radius sphere and wear amount as a result of 2a amount of shifting. Wear per unit shifting distance is calculated as $1/3\pi a^2$, by dividing $2/3\pi a^3$ to 2a. As change in the shape is permanent, Wi load is presented as Wi = H a^2 material strength and type. At the end, total wear is shown as;

$$w = \frac{\pi}{3} x \sum a^2 = \frac{1}{3\pi} x \sum \frac{\pi Wi}{H} = \frac{W}{3H} \tag{2}$$

This ($W = \sum Wi$) means the total load applied by both surfaces.

If a solid material or a solid particle removes piece by scratching or rubbing, this is defined as abrasive wear. Abrasive wear comes through as long rents on surfaces in parallel with the friction direction. A simple model based on the assumption that there is not any change on grain, it only pass through soft material by rubbing it inside is presented in Figure 2. Here, normal load is shown with W, depth on the surface caused by abrasive grain is shown with h, and cone angle is shown with υ.

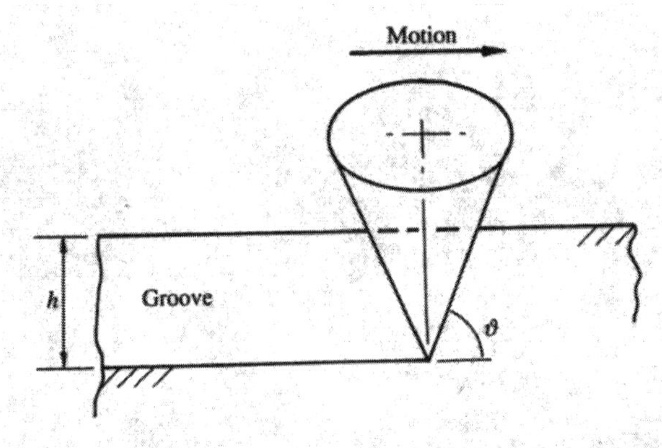

Figure 2. Movement of a cone shaped abrasive on a soft surface (Williams, 1994)

Here normal load is given as;

$$W = h^2 \text{ x } \cot\vartheta \tag{3}$$

When depth of surface caused by the grain is given as material strength, load is defined as;

$$W = \frac{\pi}{2} \text{ x } \left(h\cot\vartheta\right)^2 \text{ x } H \tag{4}$$

As a result, wear is given as this equation;

$$w = \frac{2\tan\vartheta}{\pi} \text{ x } \frac{W}{H} \tag{5}$$

If abrasive grain is prismatic instead of cone, wear becomes more complex. The structure of chip created as a result of wear is based on two angles to a great extends besides affecting forces. The first is contact angle which is the surface of abrasive grain on the side of moving angle in the direction of sliding. The second is the dihedral angle (2) which is the angle between the sides of pyramid in the direction of movement (Figure 3).

Contact angle is very important for wear, because while abrasive grain cuts chips over critical contact angles (ψ_c), it only breaks through or rubs in lower angles.

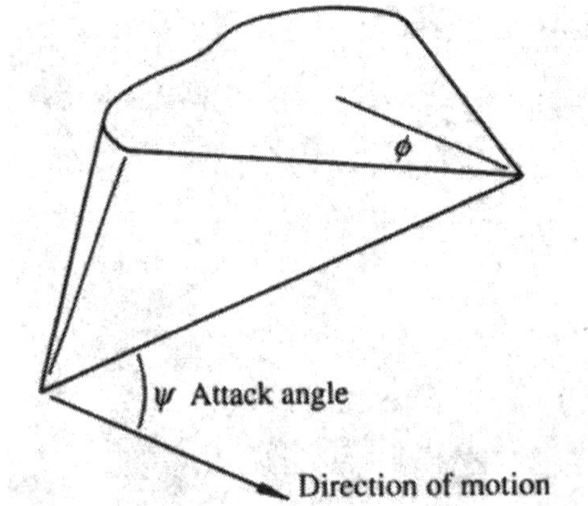

Figure 3. Geometry of prismatic abrasive grain represented with two angle ((ψ, 2ϕ) (Williams, 1994)

Dihedral angle also significantly affects the shape of chip. In very small 2 angles, abrasive grain breaks through the surface like a knife. When 2ϕ=180, it means there is a smooth surface vertical to the motion direction and this is a limiting value.

Relation between contact angle and dihedral angle developed by Kato et al. (1986) is given in Figure 4.

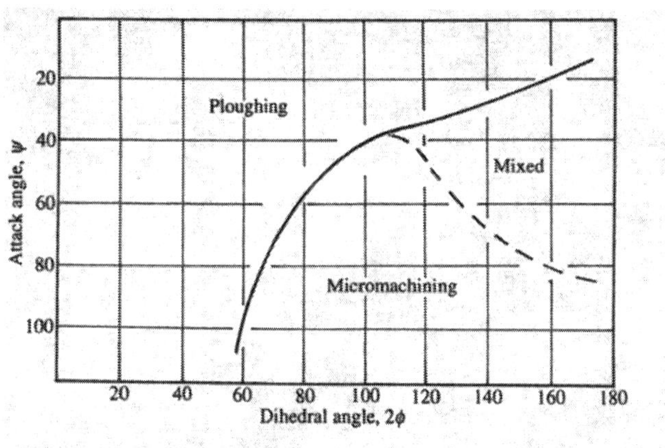

Figure 4. Relation between contact angle and dihedral angle with wear situations resulting from abrasive wear (Kato et al., 1986)

Corrosive wear occurs on surfaces that rubbed against each other with small vibrations and as a result of this, few ten micron grains are removed. Generally when irony surfaces are rubbed against one another, a reddish brown fragment is produced. This detritus is composed of solid iron oxide grains and behaves like polishing powder and make contacting surfaces smooth and shiny. This leads to the creation of a film in the shape of a protector layer on these surfaces. If wear occurs because of mechanic factors and environment involves similar atmospheric conditions, this film layer will remove and a new layer will occur as a result of re-oxidation. Grains that are formed during removal of this layer can cause abrasive wear because of their solidity. Adhesive wear can also occur as a result of friction if a part of contacting surfaces is oxidized while another part is completely non-oxidized. As a result, if this corrosion layer is continuously removing because of wear, this will have a positive effect on wear process which is named corrosive wear (Summer, 1994). Wear as a result of surface fatigue occurs generally on metal materials rolling on one another similar to the bearings. Material in contact point tightens as a result of permanent change in the shape and material embrittlement occurs. This material cracks as a result of repetitive power; they spread on the surface in time and cause breakage of material in small pieces.

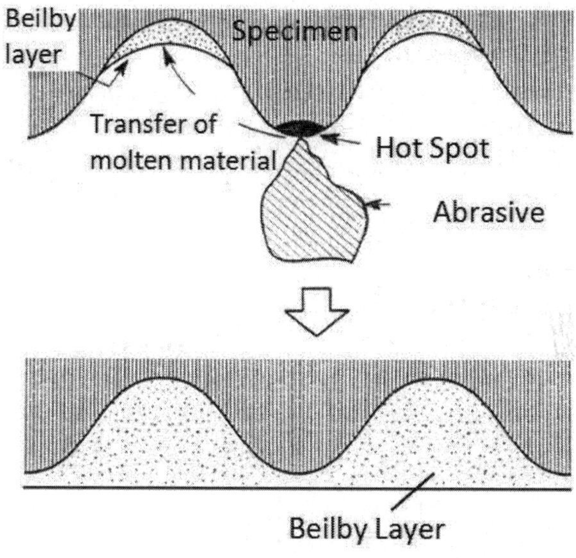

Figure 5. Beilby polishing mechanism developed by Bowden and Hughes (1937)

In the wear process resulting from surface fatigue, bigger grains remove from the surface when compared to adhesive or abrasive wear. Typical cavitations and scouring occur on these types of surfaces.

Wear mechanisms that are stated until here explain unwanted material loss on surfaces.

Abrading on the other hand, is the deliberate process of removing material from surfaces with various applications and in accordance with the purpose.

Polishing is a process of abrading and it is defined as the process of removing unevenness and visible scratches by using abrasive material (Coes, 1971). So, polished surface reflects light smoothly and in a linear way (Coes, 1971; Samuels, 1971).

According to Beilby (1921), polishing results from a smearing a material on a surface which fills the gaps on surface. Beilby (1921) stated that this material has a structure that is completely similar to amorphous and it looses crystal structure. He didn't suggest a mechanism about smearing material on these surfaces. But later on, a mechanism was developed by Bowden and Hughes (1937). These researchers determined that very high temperatures were reached in the contact points of abrasive grains, as a result of rubbing solids against each other which caused them think that heat is significant in the process of polishing (Samuels, 1971).

Unevenness on the original surface heat locally (until melting point) which is caused by friction transfers to the gaps (Figure 5). This material is transferred as a result of rapid cooling, it has an amorphous structure and constitutes Beilby layer.

On the other hand, Samuels (1971) didn't accept the existence of Beilby layer and tried to prove that Beilby layer doesn't occur with many proofs. According to Samuels (1971), there exists continuous material loss during polishing on surfaces that are physically polished. In addition, when polished surfaces are analyzed with microscope, scratches can be seen. This situation is completely opposite the existence of Beilby layer. Because Beilby stated that material fill the gaps during polishing. There shouldn't be a distinct material loss.

When physically polished surfaces are treated with acid, scratches on the surface appear. According to Samuels (1971), this situation can be explained with the deformed layer (as can be seen in Figure 6) rather than Beilby layer.

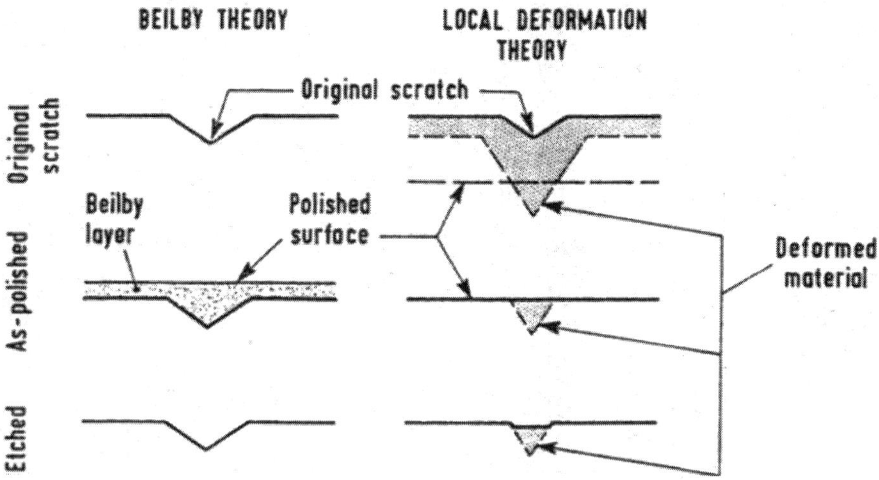

Figure 6. Comparison of Beilby and local deformation theory (Samuels, 1971)

According to Samuels (1971), one needs to explain three acceptations in order to explain physical polishing mechanism. The first is that, surfaces always have thin scratches or joint sets. The second point is that material is removed from the surface during polishing process with a constant speed. Thirdly, a layer that has permanently deformed is created. This layer is highly similar to transformed layer that is created during abrading. When grinded and polished surfaces are examined, the significant similarity between them will be seen. Scale is the only difference. Most of the researches and studies on abrading and polishing process focus on metallic materials. Studies on brittle and fragile surfaces like rock surface are very limited.

In this section, abrading and polishing processes that are based upon mechanic materials are taken into consideration. When abrading and polishing mechanisms are analyzed in these terms, it is seen that there is not a basic difference between them. By changing the abrasive material type and/or application style of gain size, abrading process can be transformed to polishing.

Explanation of abrading and polishing mechanism is possible only by revealing the type and aim of the applied process. So, mechanisms that occur at each application will be different from one another.

These applications can be lined as;

- Wear mechanism formed during the use of circular saw, grinding mills, and grinding cutting stones.
- Wear mechanism during cutting process with diamond blade saws.
- Wear mechanism during cutting process with diamond bead system
- Wear mechanism during surface polishing applications
- Wear mechanism during the use of sandpaper

Abrading processes mentioned above have significant differences in terms of basic mechanism. This is why, each one of them will be analyzed and what kind of abrading and/or polishing mechanism develops will be put forward.

3. Wear mechanism that is formed during the use of circular saw, grinding mills, grinding stones

Circular saws is the cutting tool that is used the most in cutting and sizing of natural stones that are segments containing diamonds' welded around circular metal body. Grinding mills are used for process such as cutting, graving, shaping… etc. In the abrading operation mentioned until here, the process of wear results from simple geometrical situations with a linear movement between abrasive grain and material. When grinding mills are analyzed, if abrading operation occurs on the edge of grinding mill, the situation is simple. But if abrading is on the disc, the situation is complex. In the literature, abrading mechanism that occurs here is mentioned with the word grinding. The wear that occurs here is defined as a micro-scaled grinding.

Salmon (1992) tried to make a mathematical modeling for the abrading operation on the surface of grinding mill.

Situation of grinding mill during the operation is geometrically shown in Figure 7.

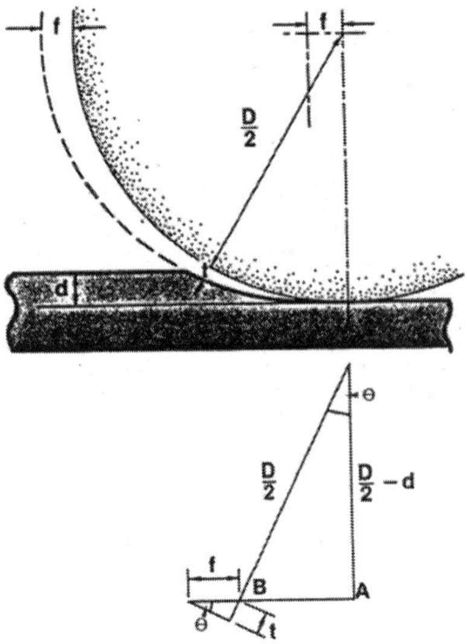

Figure 7. Geometric presentation of wear produced during grinding mill application (Salmon, 1992)

Symbols that are shown here and that will be used hereafter is shown below:

D: Diameter of grinding mill (mm)
V_s: Peripheral speed of grinding mill (m/s)
w: Angular speed of grinding mill (rad/s)
V_w: Feedrate of material
K: Number of abrasive grains along a peripheral line
C: Number of active abrasive grains per unit
t: Cutting depth of abrasive grains (mm)
l: Length of cutting trajectory
d: Cutting depth of grinding mill (mm)
b′: Theoretical width of each grain (mm)
b: Cutting width of abrasive grain (mm)

Salmon (1992) made these approaches;

Length of cutting trajectory can be determined as below:

$$l^2 = D^2 / 4 - (D/2 - d)^2 + d^2$$

$$= D.d$$

$$l = (D.d)^{0.5} \tag{6}$$

According to the geometrical structure in the figure;

$$AB = \left((D/2)^2 - (D/2-d)^2\right)^{0.5}$$

Calculated as such:

$$AB = \left(d(D-d)\right)^{0.5}$$

Cutting arc is accepted to be a straight line and the mill -along a peripheral line- that contain K amount of abrasive grain, proceeds as much as f with 1/K turn. Cutting depth of abrasive grains is calculated as:

$$t = f(Sin\theta) = f(AB)/(D/2) = 2f\left(d(D-d)\right)^{0.5}/D$$

As d, is at a small value according to D,

$$t = 2f(d/D)^{0.5} \quad f = V_w/(Kw)$$

$$t = (2V_w/Kw)(d/D)^{0.5} \ \textit{can be written} \tag{7}$$

Number of abrasive grains along a peripheral line ise determined with the equation below.

$$K = \pi.D.b'.C$$

If abrasive grain is represented as such (Figure 8);

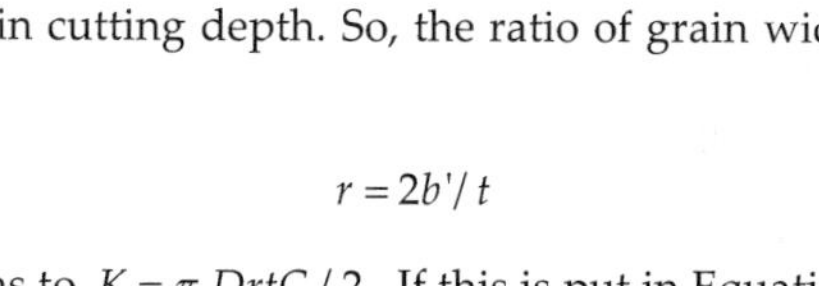

Figure 8. Frontal view of the presented grain (Salmon, 1992)

t/2 is used as average grain cutting depth. So, the ratio of grain width to cutting depth will be:

$$r = 2b'/t$$

$K = \pi.D.b'.C$ equation turns to $K = \pi.DrtC/2$. If this is put in Equation 7;

$t = (d/D)^{0.5}(4V_w/\pi DV_sCrt)$ is reached. $V_s = \pi.D.\omega$ so;

$$t^2 = (4V_w d/V_s Crl) \tag{8}$$

Cutting geometry when abrasive grain is considered, cutting geometry is presented in Figure 9.

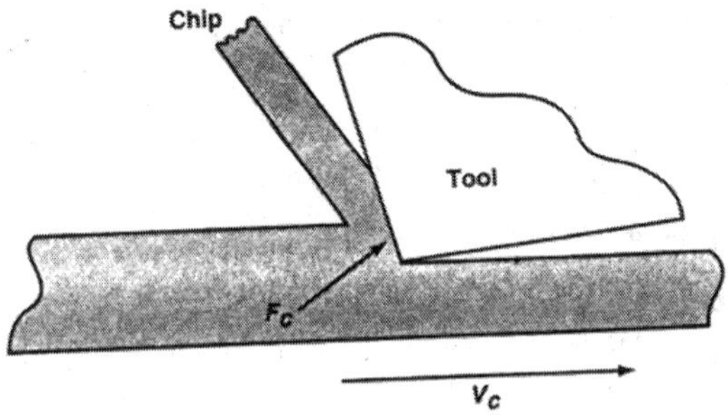

Figure 9. Cutting geometry of a abrasive grain. (Salmon, 1992)

Here;

b: Cutting width (mm)
t: Thickness of chip that is not deformed (mm)
u: Specific energy (Jmm^{-3});
For cutting operation at one point;

$$u = F_c V_c / b + V_c$$

$$\text{Cutting Force} = F_c - ubt$$

Cutting force in abrading is used as tangential force. So, equation is arranges as $u' = F + V_s / V_u bd$.

Cutting Force = $F_c u' bd V_u / V_s$

Work done in unit of time: u'Vubd

Number of abrasive grain working at unit of time: $V_s Cb$

Work done by one single grain = $u'Vud / V_s C$

Average force affecting one single grain F″ is calculated by dividing the work done by one grain into the length of cutting arc.

$$F'' = u'Vud \; / \; V_s Cl \tag{9}$$

From Equation 8:

$$t^2 = 4\; Vud / \; V_s Crl \qquad => \qquad l = 4\; Vud / \; V_s Cr\; t^2$$

When l is put into the place in Equation 9,

$$F'' = u' t^2 r \; / \; 4 \tag{10}$$

is obtained.

According to Salmon (1992), it is possible to solve wear problems in grinding mill applications and present alternative solutions.

In terms of energy need, in order to remove material from the surface, the most efficient phase is this cutting phase. Minimum specific energy is used in this way. Here, specific energy is the energy that is needed for removing unit material form the surface and unit is joule/mm^3 or Btu/in^3.

According to Salmon (1992), energy used during abrading in which chip is shaped can occur in these ways:

- Heating on working material,
- Heat occurs at grinding unit,
- Heat that occurs at chips,
- Kinetic energy at chips,
- Radiation diffused around,
- Energy spent on producing new surface,
- Residual stress in the surface and chip's lattice structure.

Another approach for grinding mills is developed by Chen and Rowe (1996). According to Chen and Rowe (1996), when a abrasive grain on the surface of a moving chip is thought; firstly, abrasive grain combines on the material with a narrow curve. In this way, more material is removed. Secondly, productive contact point of the surface of abrading changes as long as it moves on this contact arc.

So, as can be seen in Figure 10, while abrasive grain made "ploughing" at the beginning of this arc, the rest of is can make cutting.

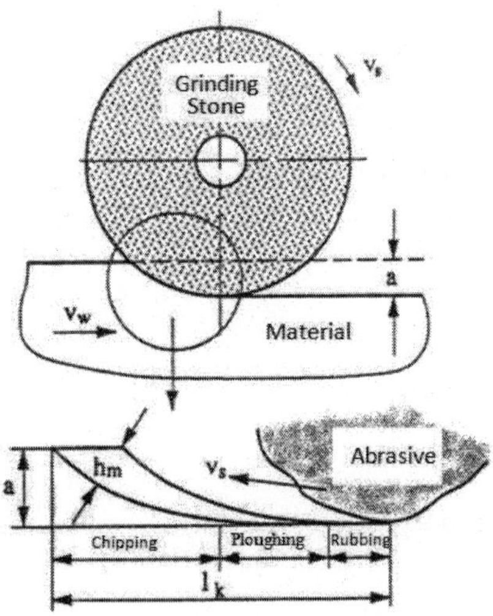

Figure 10. Phases of chip formation on the edge to grinding mill (Chen and Rowe, 1996)

In Figure 10, production of chip by grain on grinding mill during the movement of chip is seen. Cutting arc length of grain is shown with l_k, thickest chip thickness that hasn't changed shape is shown with h_m, tangential turning speed of mill is represented with V_s, progress speed of processed material is represented with V_w, cutting depth is represented with a.

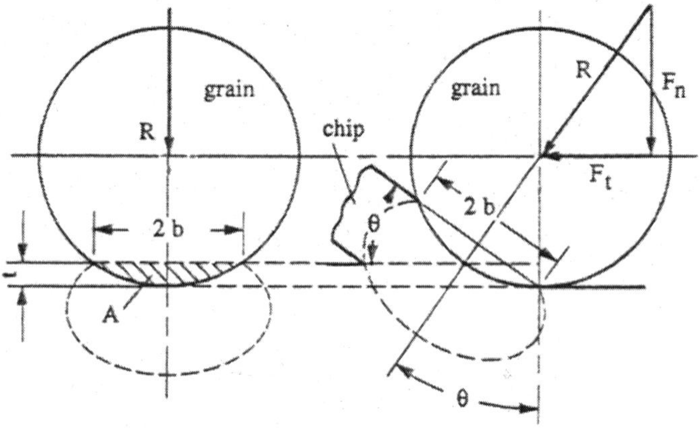

Figure 11. Behavior of circular grain in abrading (Chen and Rowe, 1996)

Movement of a circular grain on the edge of mill during abrading is shown in Figure 11. Cutting depth of grain in here is represented with t, amount of pressure with R, its horizontal component with F_t, vertical component with F_n, cross sectional area of chip that hasn't gone under any change with A, diameter of circular area which is the section of this on the surface with b, angle between power of pressure and vertical with θ. Pressure force affecting grain is represented with

$$R = \pi b \ H(C^{'} / 3) \tag{11}$$

Here C is the strain factor defined as the rate of average pressure affecting contact area to normal stress. Necessary specific energy is defined as;

$$e_c = F_t / A \tag{12}$$

When $A = \frac{4}{3} bt$, if necessary specific energy is used for cutting,

$$e_{cc} = \frac{3RSin\theta}{4bt} \tag{13}$$

When is put in the equation, specific energy is calculated as below:

$$e_{cc} = \frac{3\pi}{4}\frac{b}{t} H\left(\frac{C'}{3}\right) Sin\theta \tag{14}$$

When friction force is taken as μRCos, specific energy in friction is obtained as;

$$e_f = \frac{3\pi}{4}\mu\frac{b}{t}H\left(\frac{C'}{3}\right)Cos\theta \quad (15)$$

total specific energy for grain is formulized as below.

$$e_g = \frac{3\pi}{4}\frac{b}{t}H\left(\frac{C'}{3}\right)\left(Sin\theta + \mu Cos\theta\right) \quad (16)$$

In parallel with common use of grinding mills in the industry, there are many studies in the literature about grinding mills.

4. Wear mechanism formed in the process of cutting with diamond frame saw

Generally rock cutting mechanism is explained by the formation of indentation with plastic deformation and breaking mechanism of rock. When cutting depth of diamond is deep enough to produce visible cracks on a rock, breakages occur and chips are formed as a result of this. As can be seen schematically in Figure 12, there is a plastic deformation under the channel that is produced by the tangential movement of abrasive grain along the surface and there are two main crack systems named radial and lateral that are produced from this zone. Radial cracks are formed with wedge wear type when high normal force is used and when this force is removed, these cracks can continue to spread because of permanent tensile stress at the edge of crack. Lateral cracks start to be formed when he force is removed and can continue to spread with the effect of permanent tension (Konstanty, 2002).

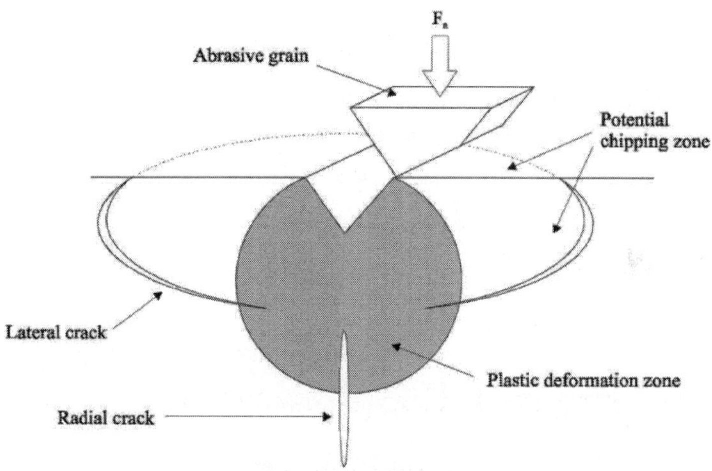

Figure 12. Schematic view of plastic deformation zone formed during cutting (Konstanty, 2002)

In the cutting process with diamond segmented blades are moved with reciprocating motion at a sinusoidal speed that is about 2 m/s. Konstanty (2002) made some researches on the cutting mechanism with diamond blade saw and defined the cutting zone during this cutting. Schematic view of cutting zone determined by Konstanty (2002) is shown in figure 13. Here, in order to make the definition, it is accepted that diamond grains in diamond zones are placed with the same protrusion height and cutting zone is the same all along the segment. But diamond grains on diamond segments have different protrusion heights. This complicates the process of defining cutting process. As can be seen in the cutting zone in figure 3.20, a pressure is produced on matrix as rock fragments accumulated in the front and behind the diamond grains can not be removed. Magnitude of pressure resulting from the wedging of rock fragments cause wear in the contact zone that is the weakest point between diamond and matrix.

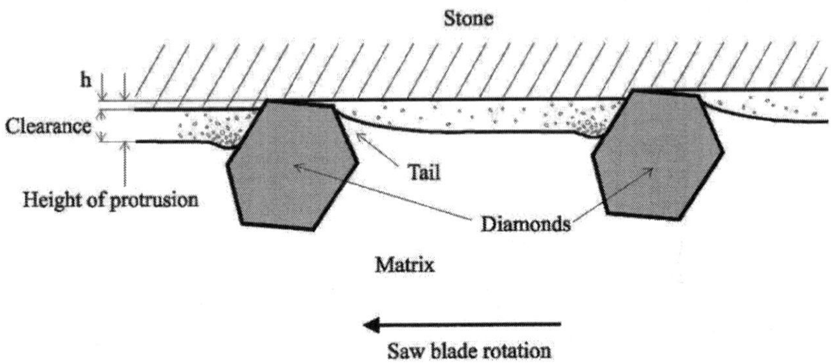

Figure 13. Schematic demonstration of cutting zone in the frame sawing system (Konstanty, 2002)

In the process of cutting with diamond blade saw, cutting of diamond segments is similar with cutting of many diamond grains. Cutting principal of diamond grain in segments are shown in Figure 14.

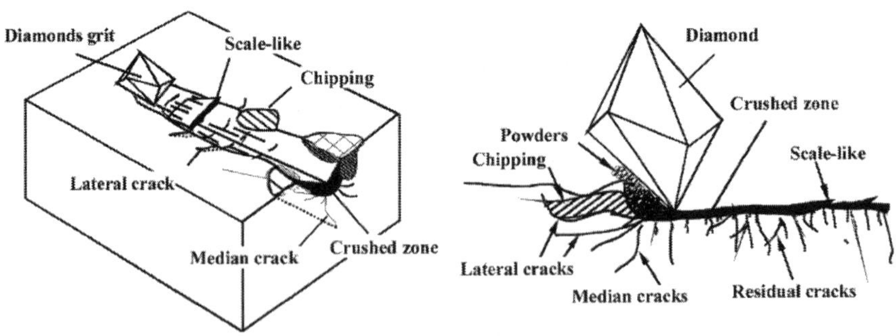

Figure 14. Cutting mechanism of marble with diamond grain (Wang and Clausen, 2002)

As can be seen in Figure 14, the main deformation of natural stone in low cutting depth is explained as plastic deformation. In parallel with the increase in cutting depth, while lateral cracks increase, plastic deformation of natural stone decreases and as a result, chip is formed. Some small lateral cracks on the surface can have a flaky structure on the base of cutting channel like a shell. Lateral cracks on different directions leave the semicircular channel behind the diamond the cuts on the surface. Plastic deformation zone stays on the base of cutting channel. Divergence on the cutting zone resulting from the increase of shearing cracks on the surface along the breaking zone seems like a continuous chip formation (Wang and Clausen, 2002).

Cutting with diamond blade saw is the continuous cutting movement of many segments on the surface of rock. Cutting with diamond segment can be defined as the cutting of a diamond cutter that cuts from many points in different cutting depths. As diamonds make chips and cuts, cracks are formed and they join and as a result of this, natural stone is broken. This situation is given in Figure 15 as stated by Wang and Clausen (2002).

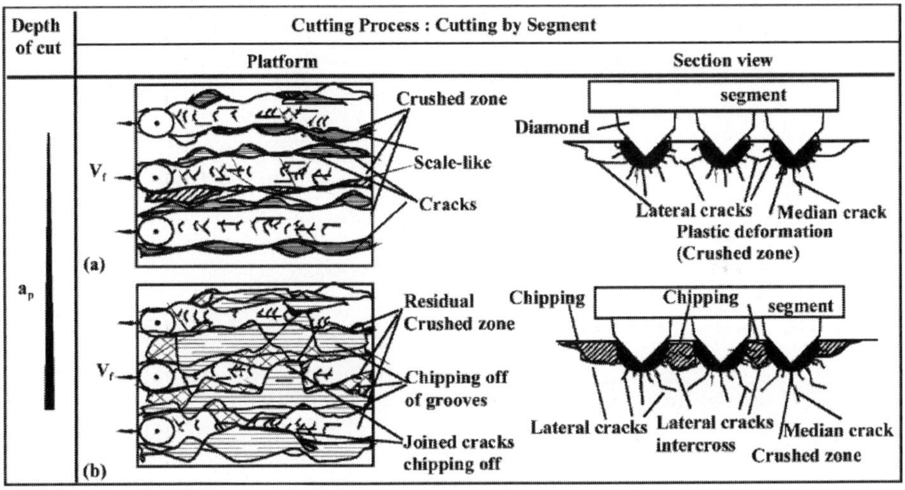

Figure 15. Cutting process of marble with diamond segments (Wang and Clausen, 2002)

Natural stone cutting mechanism with diamond blade saw system is explained as plastic deformation (breaking zone) and brittle breaking of rock. Formation of chip can be used in the explanation of cutting with frame saw system as s baseline. Plastic deformation and breaking of rock is affected from cutting conditions such as cutting depth, cooling operation, shape of cutter and aspects of rock (Konstanty, 2002; Wang and Clausen, 2002).

5. Wear mechanism formed during cutting with diamond wire system

The principle behind diamond wire cutting involves pulling a spinning, continuous loop of wire mounted with diamond bonded steel beads through the stone to provide the cutting action (Figure 16). Through the combination of the spinning wire and the constant pulling force on the wire, a path is cut through the stone. In marble quarrying through diamond wire cutting, the initial step for making a vertical cut is to drill two holes, one vertical and one horizontal, which intersect at a 90° angle. The diamond wire is then threaded through these holes, mounted around the drive wheel, and the two ends are clamped together to form a continuous loop. The drive wheel may be set at any angle, from vertical to horizontal, required to facilitate cutting.

The diamond wire is comprised of a steel cable on which small beads bonded with diamond abrasive are mounted at regular intervals with spacing material placed between the beads. The beads provide the actual cutting action in this operation. They are bonded with diamond by one of the two methods: electroplating or impregnated metal powder bonding.

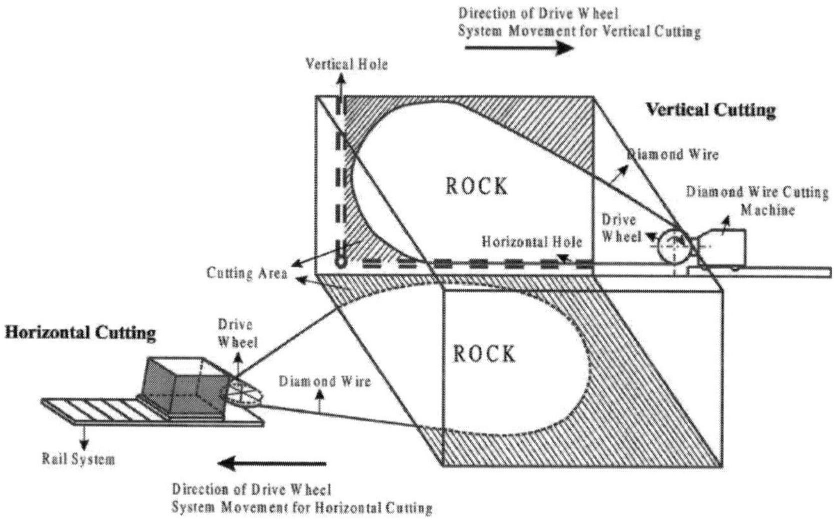

Figure 16. Situation of the diamond wire during the cutting operation (Ozcelik et al., 2002)

In the process of cutting with diamond wire system (Figure 17), diamond grains, sintered with metal powder as bead form, contact material surface similar with the circular saw and make grinding, cutting or abrading processes according to contact angle.

During cutting natural stone with diamond wire, contact angle between diamond grains and rock surface vary according to the path of steel rope (on which diamond beads are lined) in the rock rather than the diamond grains on bead.

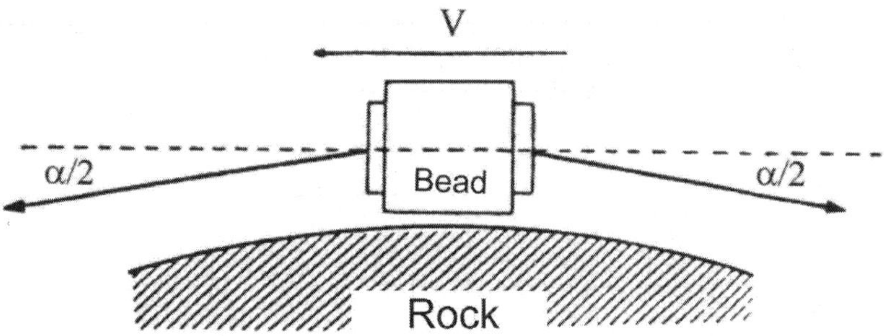

Figure 17. Schematic presentation of cutting the natural stone with diamond bead (Bortolussi et al., 1994)

6. Wear mechanism formed during surface polishing applications

This mechanism includes abrading and polishing mechanisms on grinding heads that are used for leveling and polishing of surfaces of different materials.

Although some of grinding mills are used in the literature, there are some significant differences between the grinding mechanisms of grinding mills. Abrading that is made with the help of abrasive grains on the surface of grinding heads is in fact quite similar with the abrading on sandpaper. Abrasive grains on the abrading product makes cutting, rubbing or ploughing as of their position. According to the application of abrading operation, complication of the mechanism is somewhere in between the mechanisms that are produced in sandpaper and grinding mill applications.

If grinding heads is turning around vertical axis on material surface, grains on it will make cutting, breaking through or friction during abrading. If there is the linear movement besides turning, the situation will be more complicated. Abrasive grain will be able to make these three moves during operation in different times.

In figure 18, the model that is developed by Lawn and Swain (1975) about crack movement on material surface and material removal. At the first contact point between grinding and surface, because of the applied loads high stress occurs. If the tip of the abrasive grain is perfectly sharp (namely, if radius of curvature is 0) stresses at this point will be infinite. These dense stresses relax with permanent changes in the shape and changes in density.

When applied loads reach a critical value, the middle crack shown with M start to increase because of tensile stress occur on vertical plane. In parallel with the decrease of load, middle fracture filling and when it reduced more, lateral cracks shown with L occur. These cracks are formed as there are residue elastic stresses after relaxation in contact points. Crack reaches surface with the removal of complete load and it cause wear with the breaking of material from surface.

According to Chandrasekar and Farris (1997), a few mechanisms are dominant in removing material from brittle surfaces. These are brittle break that is formed according to crack systems that is parallel and vertical to the surface and ductile cutting in the shape of chip similar to slim ribbon. The process that will occur depends on the load on abrasive grain, location and velocity of slip. Abrading process cause destruction in places close to the surface in the shape of small scaled crack, residual stress and permanent change in shape.

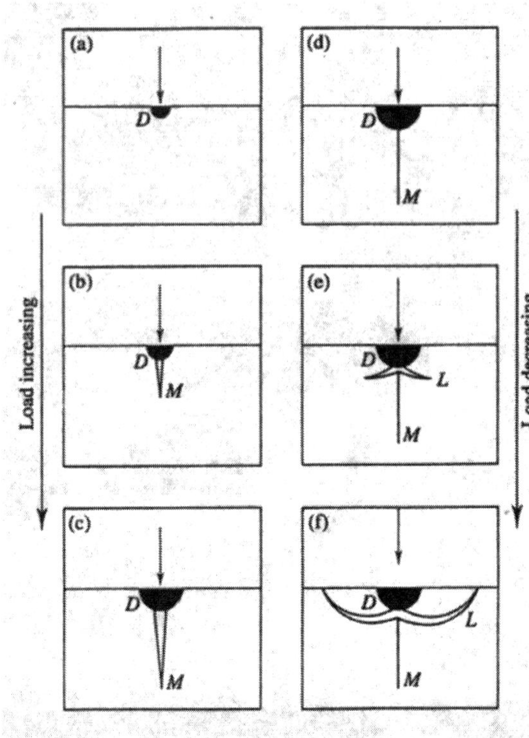

Figure 18. Formation of crack on brittle material (Lawn and Swain, 1975)

Material wear observed in the surfaces analyzed under electron microscope are in these manners; breaking of pieces by breaking of lateral cracks in parallel with the surface, big cracks resulting from breaking of grains on the surface, breakages resulting from uniting of other cracks and radial cracks and cutting movement that produce chip like metals. Formation of mechanism is proportionate to the load on abrasive grain. If load affecting abrasive grain is little, plastic micro cutting or escalloping mechanism is dominant. Surface that is formed with this process is very remarkable with it smoothness. Plastic micro-cutting movements cause creation of chip. If big loads affect discs, brittle cracks are formed on the surface. The most common types of material loss that is caused by brittle cracks are –as mentioned before- lateral crack breakings, breaking of grains on the surface and breaking of pieces from the surface in the shape of spalling. In order to understand the mechanism better, the model developed by Chandrasekar and Farris (1987) is given in Figure 19.

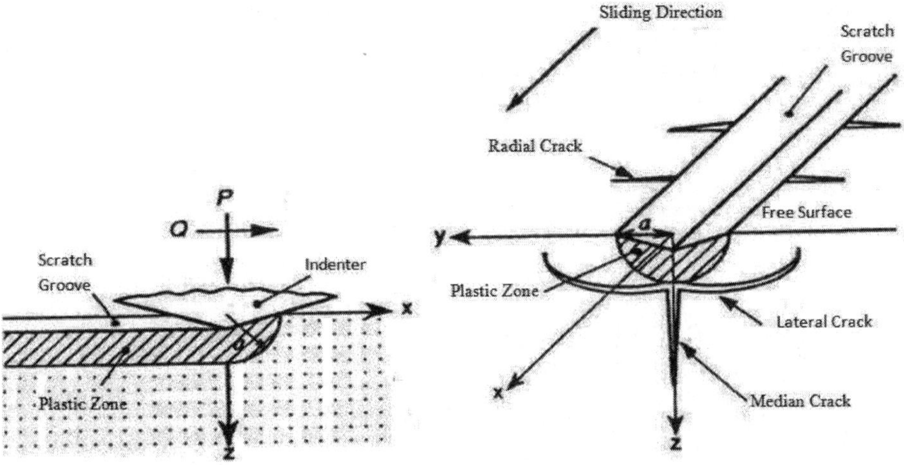

Figure 19. Slipping of grinding indenter on brittle surface and schematic demonstration of fracture after the process

Here, it is thought that one single abrasive grain slipped over the surface and groove. Normal load is very low and groove following a permanent change in the shape without breaking. It is assumed that middle cracks are vertical to the surface and the depth is in direct proportion to the size of normal power applied on abrasive grain. Middle crack starts to unite with lateral cracks in parallel with the increase in normal power. At high loads, lateral cracks are broken and cause material wear. Again at high loads, scratches break along the middle crack and material wear occurs. A similar model is developed by Regiani et al. (2000) (Figure 20).

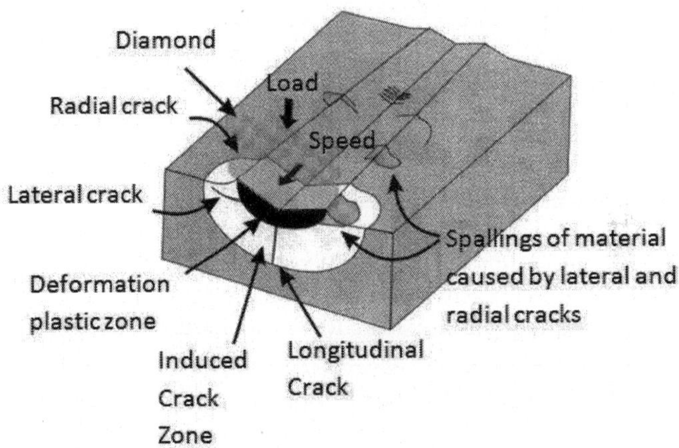

Figure 20. Cracks on the surface formed during abrasion (Regiani et al., 2000)

According to Regiani et al. (2000), basic mechanisms in material wear are; grains breaking, smashing, formation of ductile chip, and spalling. Wear of these types of materials are affected from various variables. Viscosity of used liquid, applied force on the disc, type of the disc and small scaled form of the material that is abraded are the most important of these. Small scaled form of the material has a significant effect on the development of crack that is formed as a result of abrading.

Regiani et al. (2000) stated that small scaled formation whose shape and crystal lengths of crystals that form the material are more enduring to wear than homogenous small scaled formation whose shape and whose crystal grains' lengths are similar. Again, according to Regiani et al. (2000), intrusion, gaps and crystal grain limits behave like borders for crack progress at each type of abrading process. Used liquid, size and type of disc have the secondary importance in removing material.

As a result, volumetric wear according to different operations haven't been revealed yet. Complete understanding of wear will enable the development of productive abrading processes that will create smooth surfaces.

7. Wear and polishing mechanism formed in the use of sandpaper

It is the name of grinding product formed as a result of covering abrasive grains on sandpaper, paper or on a cloth with a binding agent. Sandpaper is widely used for abrading and polishing of surfaces that are made of metal, glass, porcelain, stone, wood...etc. materials.

When glass machine profile is analyzed, it will be seen that abrading is done with abrasive grains that are lined at different positions. Effective factors in abrading process are; disc grain shape, solidity and height, its angle to the surface, applied load and form of bonding material and so the life of paper.

Contact angle of abrasive grain and surface is the most significant effective factor on wear as is given in the definition of adhesive wear.

Abrasive grains make cutting, friction and break through movements in different amounts according to the edge structure and location. Most of the studies focus on chip formation with cutting and occasions according to this.

Chip on the surface that is created by abrasive grain is given in Figure 21 in a simplified way.

In Figure 22, chip formation on brittle material is presented. Simplified chip formation model show that change of shape generally occurs in narrow space at shear plane or in an area called shear area (Figure 21). Permanent change of shape is complex in this area. But it is probably in the shape of hydrostatic component tensile type that will stay on the new surface. Namely, crack that will enable the formation of new surface is tension crack.

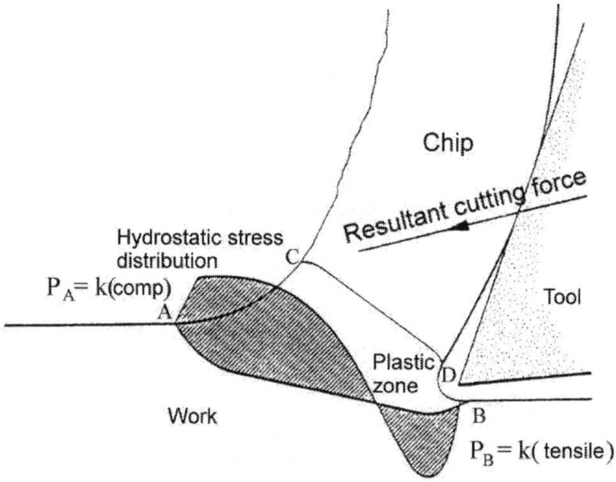

Figure 21. Chip formation model (Samuels, 1971).

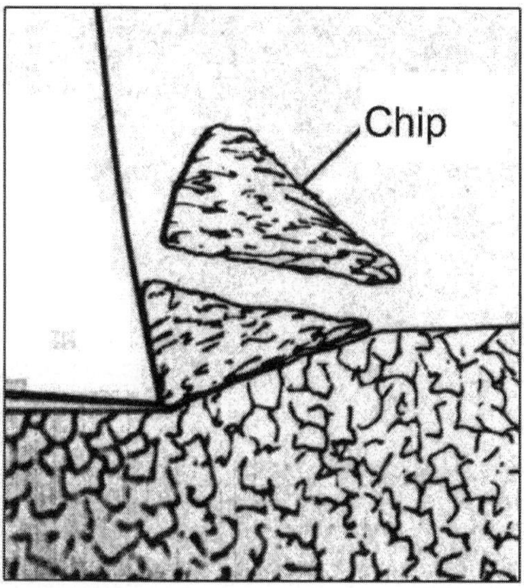

Figure 22. Discontinuous chip creation on brittle material surface (Boothroyd, 1975)

According to Boathroyd (1975), contact angle between abrasive grain and surface is very significant in the process of abrading in order to determine chip formation. on the other hand, deformation distribution of chip area is also affected from this contact angle.

As a result, there is a limiting contact angle for abrasive grains and grinding tip cuts a chip on this angle while it grooves under it. When abraded surfaces are analyzed, it was seen that very few of the scratches are formed as a result of cutting of material in the shape of chip (Samuels, 1971). Abrasive grain grooves on contact points on material surface mostly by breaking through and friction but it removes small amount of material. Very few of them create chips with grating movement and this is more effective in removing material.

Rapid material wear is ensured by applying high loads and low decreasing speed in proportion to high amounts of cutting points, namely by low wear speed of sandpaper. These factors ensure high abrading speed and prevent formation of smooth and shiny surface.

Another intended purpose of sandpaper and cloth is polishing. Mechanism in polishing process is in fact very similar to abrading. It can be said that force affecting each abrasive grain determines depth and width of scratches on the surface of basic material. So, in order to make polishing operations on abraded surfaces, very low loads should be applied on abrasive grain and abrasive grain's height should be very low. Although sandpaper is used in the first step of polishing that is made with sandpaper, polishing cloth (Figure 23) is used in the last step in order to lessen the scratch depth and form shinier surfaces.

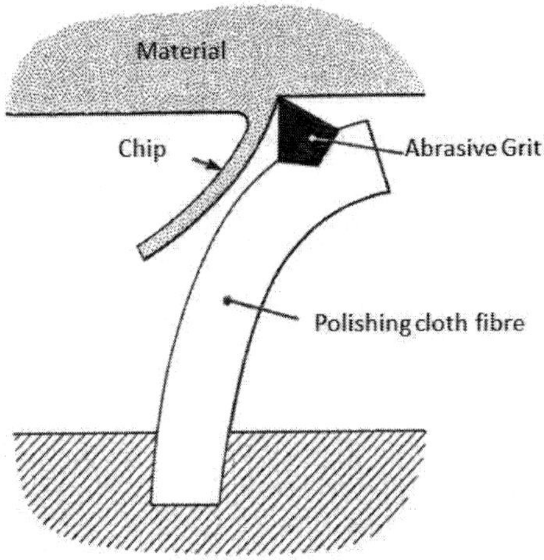

Figure 23. Mechanical polishing mechanism of abrasive grain that is clung on the polishing cloth fibre

As can be seen in Figure 23, abrasive grains in polishing cloth is inside the fibers of cloth. Grains affect on the material surface only with elastic force and ensure the creation of narrower and shallow scratches. In this way, shinier surfaces are achieved.

Author details

Irfan Celal Engin
Afyon Kocatepe University, Engineering Faculty, Department of Mining Engineering, Afyonkarahisar, Turkey

8. References

[1] Archard, J.F., 1953, Contact and Rubbing of Flat Surfaces, Journal of Applied Physics, 24, pp. 981-988

[2] Beilby, G., 1921, Aggresion and Flow of Solids, Macmillan, London.

[3] Boathroyd G., 1975, Fundamentals of Metal Machining and Machine Tools, Scripta Book Company, Washington D. C., pp. 64-106.

[4] Bortolussi, A., Ciccu, R., Manca, P.P. and Massacci, G., 1994, Computer Simulation of Diamond-Wire Cutting of Hard Rock and Abrassive Rock, IMM, Vol. 103, August, pp. A55-A128.

[5] Bowden, F.P. and Hughes, T. P., 1937, Proc. Roy. Soc., p. 575

[6] Chandrasekar, S. and Farris, T.N., 1997, Machining and Surface Finishing of Brittle Solids, Sadhana Acad. Proc. Engineering Sci.,, Vol 22, pp. 473-481

[7] Chen, X. and Rowe, W. B., 1996, Analysis and Simulation of the Grinding Process. Part 2 Mechanics of Grinding, International Journal of Machine Tools and Manufacture, Vol: 36, No: 8, pp. 883-896.

[8] Coes, L., 1971, Abrasives, Springer-Verlag, New York, p. 177

[9] Kato, K., Hokkirigawa, K., Kayabo, T. and Endo, Y., 1986, Three Dimensional Shape Effect on Abrasive Wear, Journal of Tribology, 108, pp. 346-351

[10] Konstanty, J., 2002, Theoretical Analysis of Stone Sawing With Diamonds, Journal of Materials Processing Technology, Vol. 123, pp. 146-154.

[11] Lawn, B.R. and Swain, M.V., 1975, Microfracture Beneath Point Indentations in Brittle Solids, Journal of Material Science, 10, pp. 113-122

[12] Moore, D.F., 1975, Principles and Application of Tribology, Mechanical Engineering Publication Limited, London, pp. 176-186

[13] Ozcelik Y., Kulaksiz S., Cetin M.C., 2002, Assessment of the wear of diamond beads in the cutting of different rock types by the ridge regression, Journal of Materials Processing Technology, 127 (2002), pp. 392-400.

[14] Regiani, I., Fortulan, C. A. and Purquerio, B.M., 2000, Abrasive Machining of Advanced Ceramics, IDR, 1/2000, pp. 37-65

[15] Salmon, S. C., 1992, Modern Grinding Process Technology, McGrawHill Inc. pp 83-101

[16] Samuels, L.E., 1971, Metallographic Polishing by Mechanical Methods, American Elsevier Publishing Company, Inc., Newyork, 224 p.

[17] Suh, N. P. and Saka, N., 1978, Fundamentals of Tribology, Proceedings of the International Conference on the Fundamentals of Tribology, M.I.T. press, pp. 400-405

[18] Summers, J.D., 1994, An Introductory Guide to Industrial Tribology, Mechanical Engineering Publication Limited, London, pp. 176-186

[19] Wang, C.Y. and Clausen, R., 2002, Marble Cutting with Single Point Cutting Tool and Diamond Segments, International Journal of Machine Tools & Manufacture, Vol.42, pp.1045-1054.

[20] Williams J.A., 1994, Engineering Tribology, Oxford University Press, New York, pp. 167-199.

Chapter 11

General Approach to Mechanochemistry and Its Relation to Tribochemistry

Czesław Kajdas

Additional information is available at the end of the chapter

http://dx.doi.org/10.5772/50507

1. Introduction

1.1. Compact information on the considered disciplines

Mechanochemistry and tribochemistry disciplines are of particular importance for fundamental research and tribology engineering practice. They relate to specific coupling of physical and chemical phenomena leading to initiation of heterogeneous chemical reactions due to mechanical action.

Chemical reactions in solids initiated by mechanical action had been considered for a long period of time [1]. From the viewpoint of terms, mechanochemistry and tribochemistry may be compared with terms: physical chemistry and chemical physics. In the latter term physics is first and chemistry second. In tribochemistry, friction (tribos) is the first. Since mechanics includes friction, tribochemistry should be included in mechanochemistry. Reviews on mechanochemistry [2,3] show Matthew Carey Lea as the first systematic researcher on the chemical effects of mechanical action.

Mechanochemical reactions are clearly distinct from those of thermochemical ones. To initiate thermochemical reactions an adequate heat amount has to be supplied to overcome the activation energy.

Mechanical interaction of a solid-solid enables chemical reactions to be initiated by lower activation energy than regular thermochemical reactions.

Even a very high calculated 'flash temperature' is short-lived, thus, it rather cannot initiate chemical reactions by heat. Reference [4] demonstrates that high local temperatures generated by friction at the contacts between rubbing surfaces can evolve thermionic activity, likely to be short lived and random. These localized, dynamic regions of intense thermionic emission can act as catalytic sites for chemical activity.

1.2. Definition of mechanochemistry for the present book chapter

Mechanochemistry is the science field dealing with ultra-fast chemical reactions between solids or solids and surrounding gaseous or liquid molecules under mechanical forces. There are many detailed definitions focused on selected branches of mechanochemistry. Reference [5] defines mechanochemistry as the branch of solid state chemistry where bonds are mechanically broken The bond breakage can induce electron transfers, triboelectricity (known as mechanoelectricity), and triboluminescence. These phenomena are in a branch of mechanophysics. Similarly, thermal expansion, piezoelectric effects, or compression by pressurizing might also be related to mechanophysics; details are discussed in [6].

In this chapter the following general definition, taken from [7-8] has been selected *'Mechanochemistry is a branch of chemistry which is concerned with chemical and physicochemical transformations of substances in all states of aggregation produced by the effect of mechanical energy'*. This definition was formulated 50 years ago and is accepted nowadays.

1.3. Definition of tribochemistry for the present book chapter

Similarly to mechanochemistry, also tribochemistry relates to mechanically initiated chemistry. The activation energies of tribochemical reactions are lower than those of thermochemical ones. Chemical reactions of tribological additives proceeding during the boundary lubrication (BL) process involve the formation of a film on the contact surface. BL is the condition of lubrication, in which the friction and wear between two surfaces in relative motion are determined by the properties of the surfaces and by the properties of the lubricant other than viscosity. This definition is closely related to the Hardy's first approach to the boundary lubrication process [9]. Campbell [10] emphasizes that *BL is perhaps the most confusing and complex aspect of the subject of friction and wear prevention.*

Tribochemical reactions are also distinct from those of thermochemical reactions. The same is due to heterogeneous catalysis (HetCat) and tribocatalysis [11]. Principles and applications of HetCat are compiled in [12]. To initiate thermochemical reactions heat should be supplied. There are many definitions of the tribochemistry term. Reference [13] defines tribochemistry as the chemical reactions that occur between the lubricant and the surfaces under BL conditions and stresses that the precise nature of the chemical reactions is not well understood. Book [14] just states that *tribochemistry concerns interacting chemical reactions and processes.*

The present author proposes to consider tribochemistry as a subset of mechanochemistry and thus, for this book chapter the following general definition *'Tribochemistry is a branch of chemistry dealing with the chemical and physico-chemical changes of solids due to the influence of mechanical energy'*, has been selected [7-8].

2. Flash temperature vs. thermionic emission

2.1. Frictional thermal energy

The input of thermal energy generated in a tribological system of boundary friction is lower than the output. Thus, it seems convenient to consider heat evolution also in electronic terms. Not considering any heat loss, the difference is controlled by *energy stored* in the system. Looking at the mechanical work plane proposed in [15], various portions of the work (power) include: input power, use-output power, loss-output energy rate, and a stored energy (*thermal energy transformed from mechanical work*). The energy stored (*excess energy*) points on the origin of enhanced reactivity of solids.

Mechanochemical reactions constitute a complex multistage process, which include stages involving mechanical deformation of a substance (*the supply and absorption of mechanical energy*), the primary chemical reactions, and different secondary processes [16]. It should be noted that higher reactivity of solids is of particular importance from the view-point of engineering aspects.

2.2. Brief information on the *'flash temperature'* term

The temperature rise at the peaks of the contacting asperities can be high but their duration is very short. The high order of magnitude and very short duration is due to the tiny area of contact [17]. Flash temperature is a means of accounting for the local frictional heat flux. Details related to flash temperature till 1990 are summarized in [18].

Reference [19] reviewing the present literature on flash temperature, demonstrates that for low-speed sliding, thermal effects on tribochemical reactions are negligible. Attempts to measure the contact temperature at very low sliding velocities in fretting [20], show low temperature increases and, they are well corroborated by results of theoretical calculations.

An early publication by Archard [21] deals with temperatures evolved by friction. The second part of reference [22] is on temperature distributions of friction bodies. Critical assessment of the flash-temperature concept has been presented in work [23]. It is of note at this point that frictionally generated high local temperatures can also be reflected as the thermionic emission [4].

Summarizing this information, it can be said that the flash temperature term relates to the maximum local temperature generated in a sliding contact. It occurs at areas of real contact due to the frictional heat dissipated at these spots. Theoretical study on frictional temperature rises and the flash temperature concept are assigned to Blok [24, 25] and continued by other researchers, eg. [21,23,26-27].

2.3. Thermionic emission

Work [28] shows that when the friction contact takes place at the tip of the contacting asperities, local temperatures reach significant values even in the very small time interval of

3 ms. Recent publication [29] stresses that the maximum temperature reached at a single asperity contacts corresponds to the flash temperature. As there may be many asperity contacts of different size interacting at once, there should be a distribution of such temperatures. Reference [28] also demonstrates that the high flash temperatures occur even while the overall temperature rise of the surface may be much lower. This information leads to a linkage of triboemission with flash temperature.

Now we need to come back to the present author's suggestion that flash temperature, expressed as the maximum computed friction temperature, can also be considered in terms of the thermionic emission [29]. Thereafter, flash temperatures might be expressed in the form of electronic energy. Most recently, that assumption was confirmed by examining of thermionic emission due to frictionally generated heat. The emission of electrons from a surface due to heating was investigated theoretically for sliding contacts [4]. A thermal model previously developed by Vick and Furey [30-31] for sliding contact was used to predict the temperature rise over the surface and the Richardson–Dushman equation for thermionic emission was then used to estimate the corresponding current density from the surface. The computed results demonstrate that high local temperatures generated by friction of the contacts between rubbing surfaces can activate the emission of electrons (see Figure 1). The temperature increase in Figure 1 is observable over both the actual contact area and a region immediately downstream of the contact due to energy convected by the motion of the material. The maximum temperature rise is 1003 K and the thermionic current density increase has a severe peak in the neighborhood of the maximum temperature and is almost unnoticeable elsewhere.

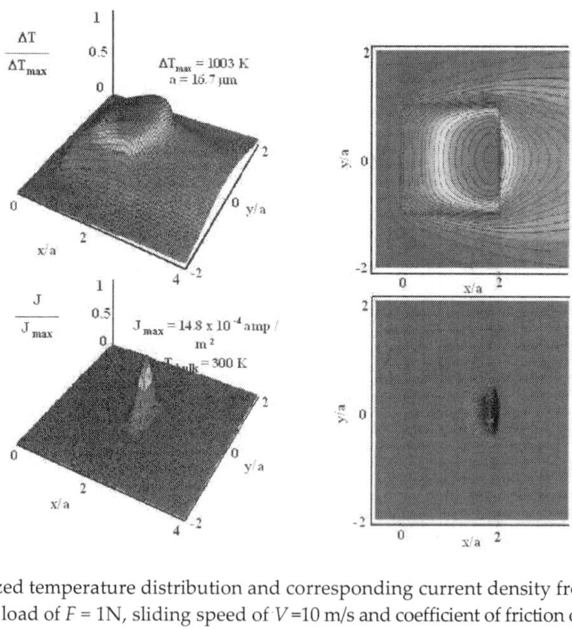

Figure 1. Normalized temperature distribution and corresponding current density from the surface of iron for an applied load of F = 1N, sliding speed of V=10 m/s and coefficient of friction of 1 [4].

This is a result of the strong exponential nonlinear relationship between absolute temperature and current density. It is evidenced by results shown in Figure 2, displaying the effect of increasing velocity or load over the narrow range of parameters. This effect is due to the strong nonlinearity and sensitivity between current density and temperature.

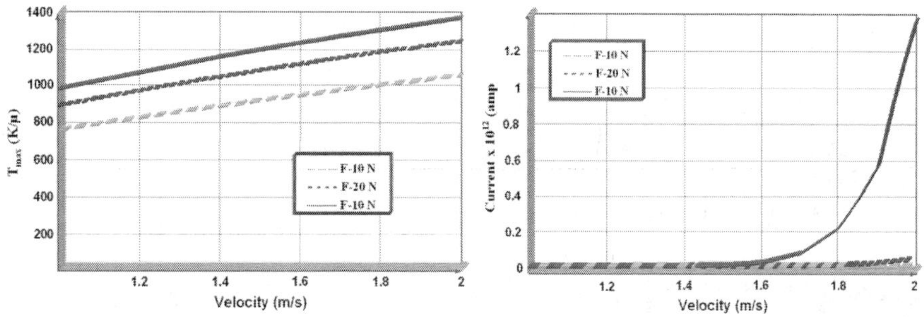

Figure 2. Effect of velocity and load on the maximum temperature divided by friction coefficient and total current from the surface of iron [4].

Two decades back tribo-stimulated and photo-stimulated exoelectron emission (TSEE, PSEE) from reactor graphite was investigated, because an increase in TSEE from graphite was due to high friction. Selective deformation of graphite fibrils was identified with transmission electron microscopy [32] and it was found that the exothermic process during the preferred orientation caused emission of tribo-stimulated electrons. PSEE from reactor graphite was also detected. New surface analysis method using TSEE has been developed to estimate the ability for metals in practical use to emit electrons. This method uses an exoelectron emission phenomenon observed only while metal surfaces are mechanically rubbed with a PTFE (polytetrafluoroethylene) rotator. Number of electrons emitted from copper and iron metals during rubbing was measured using a newly hand-made electron counting device [33]. The relationship of emitted electrons to the XPS (x-ray photoelectron spectroscopy) results of the metal surfaces also was investigated.

Thermionic emission due to frictionally generated temperatures in sliding contact can have a number of important consequences, including activation of tribochemical reactions according to the NIRAM approach and enhancement of surface reactivity [4]. Therefore similarly to the typical TSEE, generated in the sliding contact, thermionic emission (e_{th}), can be combined with the NIRAM approach in terms of enhancement of surface activity as demonstrated in Figure 3 [29].

Accordding to the NIRAM approach tribochemical reactions are initiated by both tribo (**e**) and thermionic (e_{th}) electrons of low-energy.

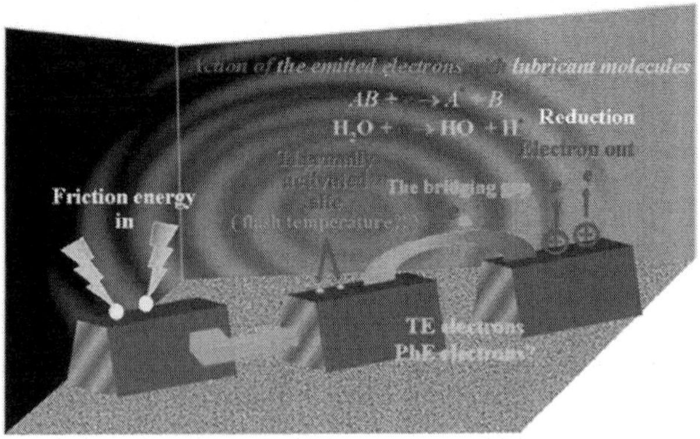

Figure 3. Initiation process of tribochemical reactions by the mechanical action.

3. Triboemission and triboplasma

3.1. Triboemission

Triboemission is defined as emission of electrons, charged particles, lattice components, photons, etc., under dry mechanical action, eg. surface damage caused by fracture processes or conditions of boundary friction. The fresh generated surface sites form a real bridge between physics and chemistry of the wear processes. Figure 4 depicts a simple idea of the exo-electron emission (EEE) process.

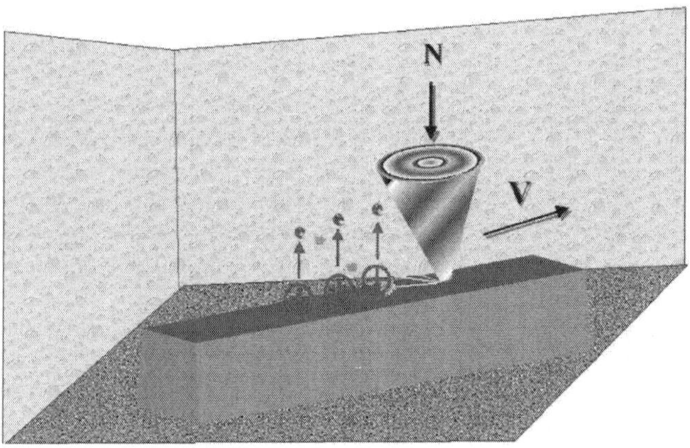

Figure 4. Exo-electron emission phenomenon.

There are many physical phenomena related to the wearing processes and mechanisms. These mechanisms are often connected with tribochemical reactions that are initiated by the surface enlargement effects. Figure 5 illustrates broadly the triboemission process.

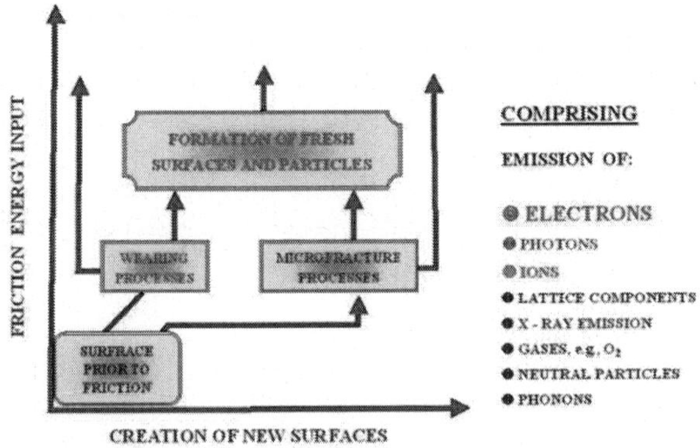

Figure 5. Physical processes evolved by friction.

Details are in reference [34], which distinguishes the following main types of triboemission phenomena: (a) emission of gas atoms and molecules, including emission of radicals and molecular clusters, (b) emission of electromagnetic radiation, (c) emission of electrons, (d) emission of ions, (e) emission of magnetic field, (f) emission of electric field including emission of electric charges and generation of tribocurrents, (g) emission of noise, vibration and acoustic emission, (h) heat evolvement.

Emission of gas atoms and molecules at friction results from the competition gas release and gas adsorption processes. When at certain sliding conditions, the rate of gas adsorption exceeds the rate of gas release, total emission rate becomes negative. Such emission of negative rate has been called *anti-emission.* The triboemission phenomena are classified into two classes by physical nature: emission of particles ('corpuscular'), and emission of energy, as shown in Figure 6 [29,34].

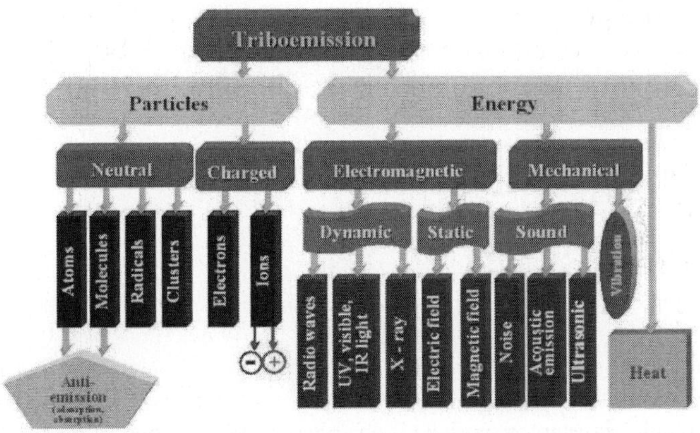

Figure 6. Classification of triboemission phenomena by physical nature.

The particle emission includes neutral particles (atoms, molecules, radicals and clusters) and charged particles (electrons, negative ions and positive ions). Three main types of energy emission encompass: electromagnetic energy, mechanical energy and heat. Mechanical energy includes mechanical oscillations of various frequency ranges, i.e. vibration, noise, ultrasonic emission, acoustic emission, etc. Electromagnetic emission can be classified into static and dynamic. Static emission includes static electric and magnetic fields, while dynamic emission includes electromagnetic waves of various wavelengths, i.e. radio waves; IR, visible, UV light; and X-rays [34]. Significant research in the field of triboemission was carried out by Nakayama et al. [35]. Kim et al. [36] investigated electron and photon evolvement (phE) from magnesia (MgO) under friction with diamond; they found that during indentation, prior to any fracture of MgO, only phE was observed. The relative intensities of these signals can therefore be used to follow the progress and extent of plastic deformation and fracture during wear on millisecond time scales [36].

The work of Molina et al. [37-38] characterized triboemission of electrons from diamond-on-alumina, diamond-on-sapphire, alumina-on-alumina, and diamondon-aluminum. The three ceramic-materials consistently showed burst-type negatively charged triboemission during contact at constant load and speed, while the aluminum system produced no significant emission. Decaying emission after contact ceased also was detected from the three ceramic systems for durations exceeding the minute-range [39]. For the cases of diamond-on-alumina and diamond-on-sapphire, energy spectrometry showed that a large fraction of the triboemitted negative charges were of low-energy (eg. 1–5 eV). This finding was of significant importance because in the NIRAM approach, it was hypothesized that the energy level of triboelectrons to initiate tribochemical reaction should be 1–4 eV [40].

Interestingly, an early work [41] demonstrated that electrons of very low energy (EEE) can be produced from solids by mechanical deformation, X-ray irradiation and chemical reaction; if the excitation is not great enough to produce normal electron emission, steady EEE can be produced by steady excitation. Another early work [42] investigated EEE from aluminum abraded in different atmospheres and found that under high and ultra-high vacuum conditions, such electron emissions is associated with a shift of photo-electric threshold dependent upon the residual gas pressure to which the surface was exposed. It was evidenced that the initial growth stage of emission is due to the adsorption of water vapor, the subsequent decay being associated with oxidation [42]. EEE process is also combined with triboplasma. Extensive review paper [43] shows exo-emission as a sensitive method for the monitoring of the initial stages in the fragmentation of polymeric materials subjected to mechanical action, and demonstrates that the electrons emitted when solids (dielectric materials, metals, and polymers) are subjected to mechanical influences are capable of inducing the dissociative ionization of the water molecules adsorbed on the surfaces. The studies reviewed in [43] demonstrate unambiguously the interrelation between the mechanoemission phenomena and the mechanochemical processes. A correlation between the electron emission phenomenon and mechano-chemical processes in solids is also presented by Khrustalev [44].

3.2. Triboplasma

Elastic deformation is the first stage of the mechanical energy interaction between solids and leads to a change of the bond distances in the affected solid. Actually, it concerns mechanical energy transmission to solids. By and large, there are few processes for the transmission of mechanical energy by impact treatment. Heinicke in the Tribochemistry book [8] states the following: *'Immediately at the commencement of a grain colliding at high velocity with a solid surface it comes to a quasi-adiabatic energy accumulation and to the formation of an "energy bubble" at the point of action in the sub-microscopic deformation zone'.* All these specific aspects are summarized in the phenomenological Magma-Plasma Model (MPM) [7,8]. As the MPM process is combined with the EEE process, it is of very significant importance for both mechanochemistry and tribochemistry. Figure 7 depicts the Magma-Plasma Model.

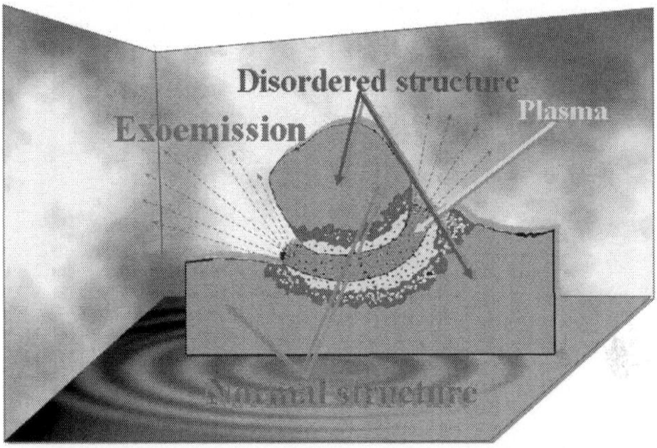

Figure 7. The Magma-Plasma Model (MPM) for the impact stress of flying grain.

The short life of the triboplasma causes no Maxwell-Boltzmann distribution, thus an equilibrium temperature cannot be given and the chemical process taking place in the excitation phase cannot be described by the laws of thermodynamics [8]. The highest stage of energy excitation changes dynamically into the next stage characterized by the relaxation of the plasma states and is termed as "edge-plasma" and "post-plasma". Work [45] describes the energy dissipation on solids activated by impact along with adequate discussion.

Mechanical forces make the atoms leave their equilibrium positions due to lattice vibration, alter bond lengths and angles of their atomic arrangements leading to electronically excited states (see Figures 3,5,7). The energy field incurred by the tribophysical effects, for example the EEE process, triboluminescence, triboplasma, crystal defects etc., initiates specific tribochemical reactions.

Mechanochemistry in terms of processes being triggered in the solid state chemistry, due to the application of mechanical energy, is extremely complex. Tribochemistry as its branch is even more entangled. Our present knowledge shows that mechanochemistry is also well combined with nanoscience [46].

4. NIRAM and HSAB (hard and soft acids and bases) and catalytic approaches

4.1. Basic information on NIRAM

In brief the NIRAM approach comprises the following major steps.

- Low-energy electron emission and generation of positively charged spots;
- Interaction of the emitted electrons with the lubricant molecules producing negative ions and radicals on tops of the rubbing surfaces;
- Reaction of negative ions with positively charged sites of friction surfaces;
- Other reactions, producing organometallic or inorganic film, which protects the rubbing surfaces from wear; if the shear strength is high chemical bonds of organometallic compounds are cleaved resulting in producing inorganic films and further radicals;
- Eventual destruction of protective layer caused by wear, followed by electron emission and subsequent formation of a new protective film.

Figure 8 illustrates the NIRAM reaction cycle. Most recent review [11] is on the NIRAM approach and shows its interrelationship with tribocatalysis. The application of the NIRAM approach explains the role of exoelectrons in some relevant tribochemical reactions detailed in [29].

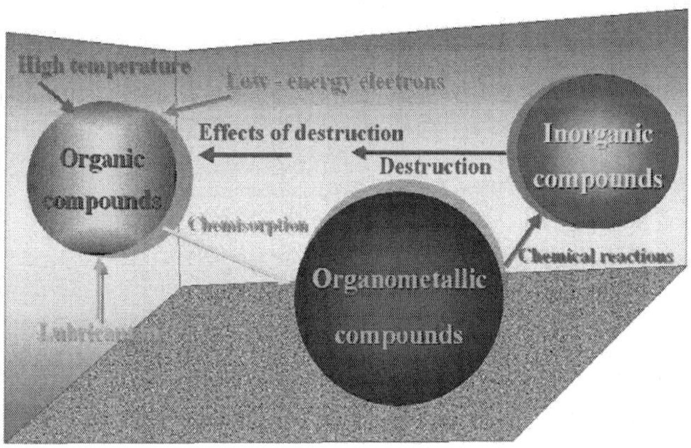

Figure 8. Reaction cycle of lubricant components on solid contacts during friction.

In summary, this boundary lubrication model proposes the formation of protective organometallic and inorganic layers on rubbing surfaces. The initiation reaction process is due to mechanical action evolving emission of low-energy electrons (1-4 eV) combined with flash temperature expressed in the thermionic emission (see Figure 1).

4.2. Examples of NIRAM applications

4.2.1. Traditional approach to mechanisms of AW and EP lubricant additives

Tribochemistry of lubricating oils is overviewed in Pawlak's book [47]. Important and interesting approach to oil formulations and complex lubrication processes is assigned to inverse micelle. Figure 9 illustrates the inverse micelles involvement to interactions of base oils with major engine oil additives. The nature of the tribochemical film is the key to better understand the mechano-chemical processes that give rise to chemical films separating mating solid elements and thereby reducing wear and seizure.

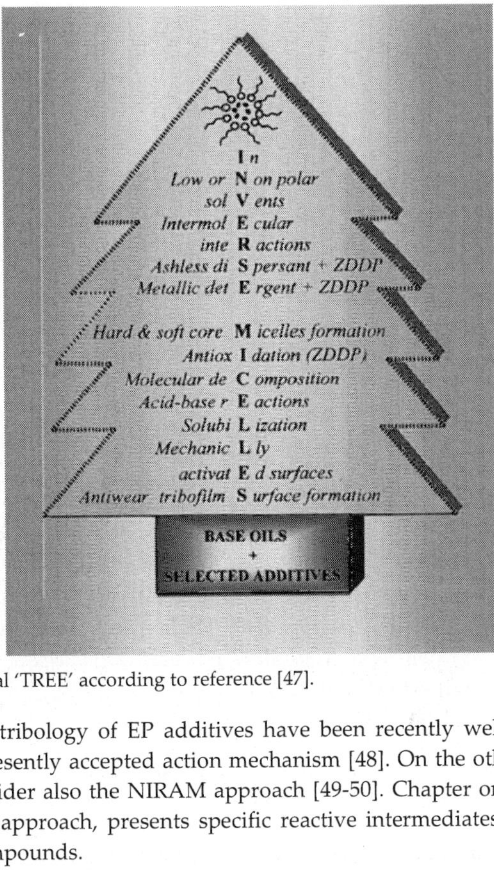

Figure 9. Tribochemical 'TREE' according to reference [47].

The chemistry and tribology of EP additives have been recently well reviewed from the viewpoint of the presently accepted action mechanism [48]. On the other hand, chapters in the same book consider also the NIRAM approach [49-50]. Chapter on *'Tribochemistry'* [51] details the NIRAM approach, presents specific reactive intermediates and mechanisms of selected organic compounds.

4.2.2. Examples of NIRAM controlled tribochemical reactions

Work [52] investigated tribochemical reactions of carboxylic acids under boundary lubrication conditions and it was found that, apart from regular salt (monodentate carboxylate group), salts with double bond in α, β position and chelating symmetric bidentate carboxylate group are formed. Figure 10 presents the proposed structure of iron salt with double bonding. These types of compounds are not produced under static conditions. It was found that bidentate surface configuration increased after the occurrence of the tribochemical reactions induced by surface rubbing. This proves this finding which was accounted for by the NIRAM approach. The new finding also clearly demonstrates the difference between tribochemical and thermochemical reactions. This finding is evidenced by other research results [53] on tribochemical and thermochemical reactions of stearic acid adsorbed on a copper surface.

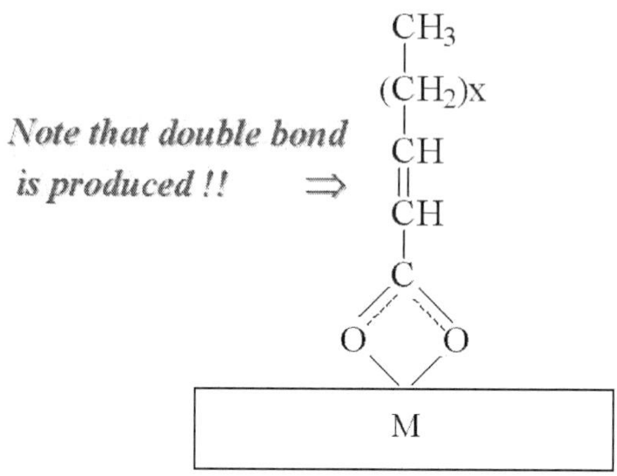

Figure 10. Chelating bidentate structure formed from caprylic acid [52].

Work [54], investigated changes of hexadecane under boundary lubrication conditions considering both (i) the chemical transformation of the bulk lubricant and (ii) the chemistry of products generated in wear tracks. Detailed analysis of hexadecane after friction tests showed that during the friction process aldehydes, alcohols and carboxylic acids are produced. This finding clearly indicates that tribochemical reactions cause significant changes of the apparently non-reactive paraffin hydrocarbon. The reaction process initiated by the frictional energy is in line with the NIRAM concept.

Regular monoester ester hydrolysis process produces alcohol and carboxylic acid. Work [55] aimed at checking role of hydrolytic reaction for the soap formation mechanism from esters. It was tried to find if typical conditions under boundary lubrication cause the hydrolysis of esters dissolved in hexadecane. Figure 11 reflects a ball wear reduction versus the additive concentration in hexadecane. These results enabled to state that the hypothesis saying *'during the friction process under boundary lubrication conditions lubricated by aliphatic esters, the ester hydrolysis process cannot proceed without an adequate catalyst'* is well substantiated. Thus, it is possible to conclude that the soap formation mechanism from esters under boundary lubrication conditions is controlled by the NIRAM approach [56].

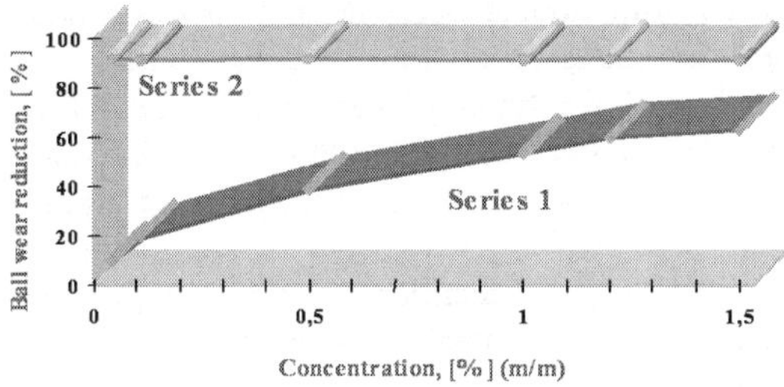

Figure 11. Influence of the solution concentration of octadecyl palmitate (Series 1) and equimolar mixture of palmitic acid and octadecanol (Series 2) in hexadecane on the ball wear [55].

The ester reactive intermediates produced via the dissociative electron attachment, showing two types of C–O bond cleavage, are presented in Figure 12. The first bond cleavage type produces carboxylate anion ($RCOO^-$) and the free radical R•. The second bond cleavage generates the alkoxide anion (RO^-) and the [R–C=O]• free radical that undergoes further reactions. To produce free radicals and thereby initiate the free radical chain reaction process either heat or catalyst is needed. Therefore, an electron attachment in this case acts as a catalyst. The exoelectron dissociative attachment to an ester molecule yielding two types of negative ions is clearly evidenced by the electron attachment mass spectrographic results [57].

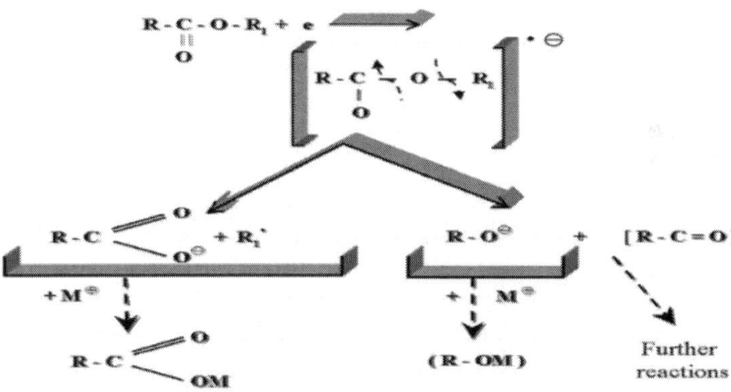

Figure 12. Ester reactive intermediates generated by electron attachment.

4.3. Basic information on NIRAM-HSAB theory

Gilman [58] emphasizes that mechanochemical effects have often been attributed to strain energy assisting thermal energy. When covalent bonds are bent or sheared, the energies of their highest occupied molecular orbitals (HOMO) are raised, whereas the energies of their lowest unoccupied molecular orbitals (LOMO) are lowered and, the gap between levels determining a bond's stability is decreased [59]. If the strain becomes big enough to close the gap, the bonding electrons can move freely, and, the reaction can take place immediately. The simplest way of illustrating that phenomenon is based on a generalized NIRAM–HSAB theory and provides its possible application for accounting for some tribochemical processes under boundary lubrication conditions. Figure 13 [11] depicts the model.

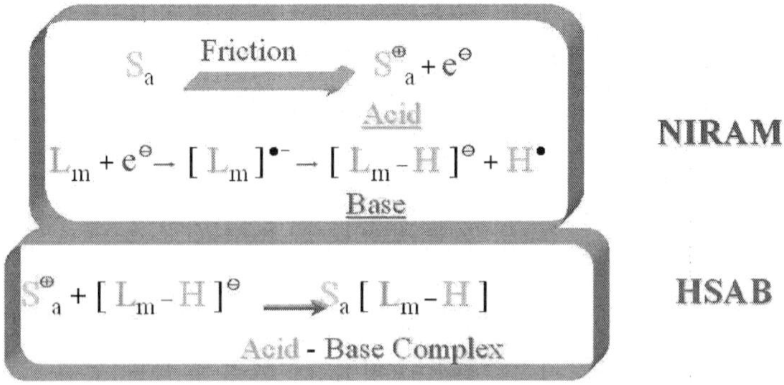

Figure 13. NIRAM–HSAB lubrication mechanism approach: S_a, tribological microsurface area; $\mathbf{L_m}$, lubricant molecule; and e⊖ is a low-energy electron emitted under boundary lubrication conditions

Presently this model has been applied to account for very complex tribochemistry of silicon nitride (Si_3N_4) [60]. The tribochemical reaction pathway of Si_3N_4 was accounted for in terms of the NIRAM–HSAB theory, in which tribo-electrons play an important role to decrease the activation energy. This may explain the reason why some products can be formed only by friction such as the tetrasiliconalkoxide obtained in lubrication with alcohols. Tribochemical wear provides flat surfaces, and decreases stresses since insoluble tribo-products act as lubricants by forming protective films, such as hydrated silicon oxides when water is present, and silicon alkoxide polymers in the case of alcohols. Figure 14 accounts for tribochemistry of Si_3N_4 with water.

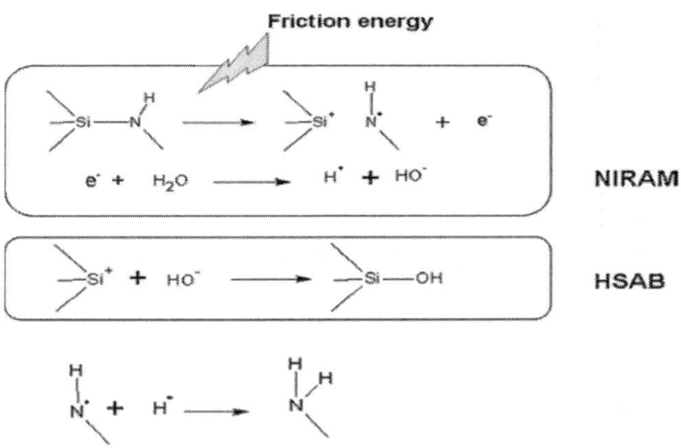

Figure 14. Interpretation of the tribochemical reaction of water with silicon nitride based on NIRAM–HSAB theory [60].

Applying the NIRAM–HSAB approach, the formation of silicon compounds from Si_3N_4 lubricated by alcohols is represented in Figure 15, which follows the scheme of Figure 14. The enhanced reactivity of silicon nitride under friction relates to active sites such as dangling bonds, and the action of tribo- and/or thermionic emissions. It should be noted that, according to Figure 16, the intermediate is the anion RO^-, which is formed also during polymerization of the silicon alkoxide [61-62], thus further increasing the rate of the tribochemical reaction.

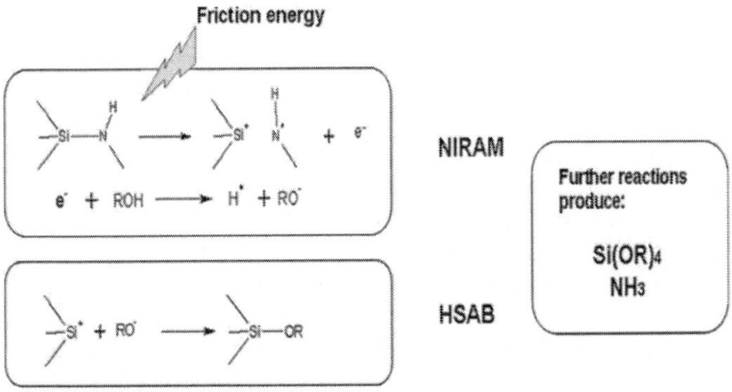

Figure 15. Interpretation of the tribochemical reaction of alcohols with Si3N4 based on NIRAM–HSAB theory [60].

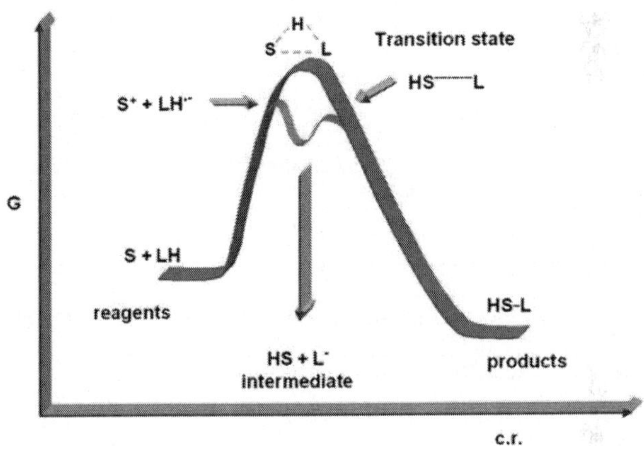

Figure 16. Free energy Gibbs (G) curve as function of the coordinate of reaction (c.r.). S is an active surface site, LH - lubricant molecule (H is hydrogen, L is the hydrocarbon branch of the lubricant) [60].

In the intermediate depicted in Figure 16, electron and proton are exchanged between the two reagents S and L. This exchange mechanism is based on the subatomic particles involved in chemistry and, it is clear relationship between the intermediate and transition state reflected in the product formed. So that the tribochemical reaction pathway of silicon nitride has been accounted for in terms of the NIRAM–HSAB theory, in which tribo-electrons play an important part to decrease the activation energy or increase the reaction rate. This may explain the reason why some products can be formed only by friction such as the tetra silicon alkoxide obtained in lubrication with alcohols.

4.4. Catalysis and tribocatalysis

Catalysis is the phenomenon of a catalyst action and the catalyst is a substance that increases the rate at which a chemical system approaches equilibrium, without being consumed in the process [12]. To initiate thermochemical reactions, the reaction system temperature should be increased to overcome the activation energy barrier (see Figure 16). The same is due to the catalytic process, but the catalyst lowers the reaction activation energy. Usually, it is demonstrated by the difference of the activation energy (E_a). Considering tribocatalytic reaction as the tribochemical one enhanced by the rubbing of catalyst, most recently it was demonstrated that in the tribocatalytic ethylene oxidation, the activation energy (E_a) with friction was less than 2% of the thermochemical reaction [63].

Activation energy lowering is a fundamental principle of catalysis and it applies to all forms of catalysis. For catalytic process to occur, a chemical interaction between catalyst and the reactant-product system is necessary, however this interaction should not change the chemical nature of the catalyst. With a catalyst, the energy required to go into the transition state decreases, thereby decreasing the energy required to trigger the reaction process. The rates of chemical reactions increase as temperature increases. Chemical reactions have rate constants approximated by the Arrhenius equation

$$k = A\ \exp^{(-Ea/RT)} \tag{1}$$

where **A** is the pre-exponential factor for the reaction. Heterogeneous catalysts provide a surface for the chemical reaction. Most heterogeneous catalysts are solids that act on substrates in a liquid or gaseous reaction mixture.

Reference [64] reviews and discusses the effect of mechanochemical activation on the catalytic properties of different systems. It is noted that the activity of a catalyst with defects is somewhat higher than its activity in the equilibrium state, emphasizing that the degree of increase in activity depends on the amount of the excess energy stored. It is stressed that the energy stored in defects influences the catalytic properties through the variation of thermodynamic potentials.

Practically all types of chemical reactions are accompanied by a change in energy. Some of them release energy to their surroundings mostly in the form of heat and thus are called exothermic. Conversely, some reactions need to consume heat from their surroundings to proceed. These reactions are called endothermic. Reactions that proceed immediately when two substances are mixed together are called spontaneous reactions. The application of mechanical energy associated with friction releases physical processes that can be the cause of tribochemical reactions of solids with lubricant molecules. Figure 17 demonstrates a general approach to physical and chemical events relating to boundary lubrication conditions.

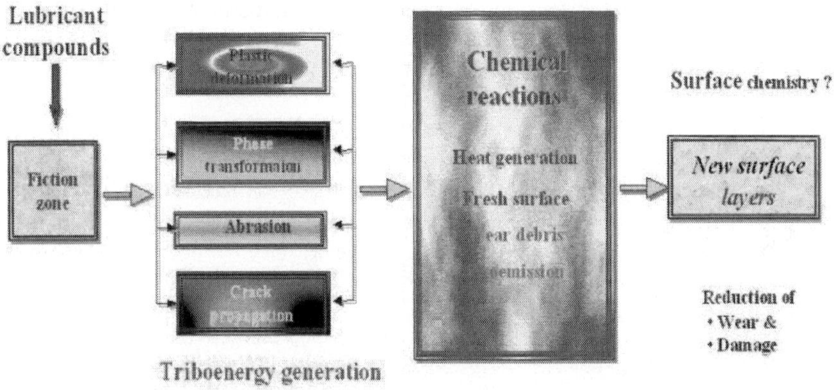

Figure 17. Major physical and chemical events in the boundary lubrication contact.

The common denominator of these reactions is that they are triggered by low-energy electrons. This statement is relevant to hypothesis of the present author saying that the intermediate reactive species of both mechanochemical and tribochemical reactions are produced by the same mechanism. Mechanolysis is very special branch of mechanochemistry, particularly from the view-point of water splitting technology.

5. History and present state of the mechanochemistry and its relation to tribochemistry

5.1. Summary of historical background based on Takacs's work

The feasibility that chemical reactions in solids can be initiated by mechanical deformation and/or tribological contact, had been considered for almost 120 years back. At that time M. Carey Lea [1] wrote *'Mechanical force can bring about reactions which require expenditure of energy, which energy is supplied by mechanical force precisely in the same way that light, heat, and electricity supply energy in the endothermic changes which they bring about'.*

Typical historical reviews of mechanochemistry by Takacs [2-3] have presented Matthew Carey Lea as the first systematic researcher on the chemical effects of mechanical action. Mechanical energy triggering chemical reactions had been of particular significance in development of mechanochemistry, known for over two centuries.

Early twentieth century (1919) Ostwald considered mechanical energy influence on chemical reactions and coined the term 'mechanochemistry'. However, at that time Ostwald was not able to say anything on the independent character and the importance of this field of chemistry. Another detailed review paper [6] on mechanochemistry deals also with its historical development. Superb experimental historical achievement of Carey Lea relates to the cinnabar (HgS) decomposition by trituration in a copper mortar with a copper pestle to produce Hg element and thereby to combine mechanochemistry with tribochemistry, due to the fact that trituration relates to friction 'tribos'. This clear description of the mechanochemical action on initiation of chemical processes is described in Takac's papers [3] as follows 'Native cinnabar (mercuric sulfide, HgS) was rubbed with vinegar in a copper mortar with a copper pestle yielding the liquid metal.' According to Tackacs [3] the source of such information is assigned to "Theophrastus' History of Stones" [65], a 1774 English translation of Theophrastus' Greek original written at the end of the 4th century B.C. It is the earliest preserved text on any subject related to chemistry or metallurgy. Thereafter, the sentence on the preparation of mercury is very probably the earliest reference to any mechanochemical reaction, extending the documented history of the process by two and half millennia [3].

5.2. Action mechanism of the first mechanochemical discovery

Another paper [66] reviews widely and precisely mechanochemistry of solids. Describing reaction of mechanochemical synthesis a kind of conclusion is clearly expressed *"One of the key problems to be solved in this area is what is the start, or trigger, of the self-propagating high-temperature synthesis process? Is the Joule heat or the formation of contacts between the particles, which would be sufficient for self-heating in the contact zone to transfer the process into the regime of self-propagating high-temperature synthesis".*

This Section demonstrates how the NIRAM approach might be applied to account for the mechanism of cinnabar (HgS) decomposition to produce Hg element and thereby to interconnect mechanochemistry with tribochemistry. It is hypothesised that both typical mechanochemical and tribochemical reaction processes are mostly triggered by triboemitted negative particles. The proposed decomposition mechanism encompasses three major steps. Mechanical action emits low-energy electrons. The emitted electrons interact with Hg=S to produce negative-ion-radical (NIR) reactive species. NIR reacting with O_2 produces unstable $HgSO_2$ which decomposes to metallic Hg and SO_2. In brief summary, the NIRAM approach can demonstrate the first feasible mechanism of the cinnabar mechanochemical decomposition to produce Hg element and thereby to interconnect mechanochemistry with tribochemistry.

Now we need to consider the mechanism of cinnabar (HgS) decomposition by trituration in a copper mortar with a copper pestle to produce Hg element and thereby to more interconnect mechanochemistry with tribochemistry, due to the fact that trituration relates to friction 'tribos'. In this case also the reaction of HgS with copper (Cu) is taken into account. The whole reaction is:

$$HgS + Cu \rightarrow Hg + CuS \tag{2}$$

The reaction might proceed via the negative-ion-radical reactive intermediate

$$^{\bullet}Hg - S^{\ominus} \quad (3)$$

The intermediate interacting with copper positively charged sites can produce CuS and Hg. It is assumed that during the trituration in a copper mortar with a copper pestle, electron (e⊖) is emitted and positively charged copper atom (**Cu**$^{\oplus}$) is produced. Thus, the NIRAM based reactions take place:

$$^{\bullet}Hg - S^{\ominus} + Cu^{\oplus} \longrightarrow {}^{\bullet}Hg - S - Cu \quad (4)$$

$$^{\bullet}Hg - S - Cu + {}^{\bullet}Hg - S - Cu \longrightarrow 2Hg + 2CuS \quad (5)$$

Please note that here we have to take into account not only CuS but also 2Cu2S:

$$Cu - S - S - Cu \quad (6)$$

other considered compounds might include Cu2S and CuS2.

5.3. Comparison of the NIRAM approach with the generation process of mechano-anion-radicals (MARs) in polymers

The NIRAM concept is based on the hypothesis that low energy electrons (1 to 4 eV), emitted from rubbing surfaces, can be the key factor in some tribochemical reactions [11,29,40,51].The concept of the MARs generation process in polymers is based on the polymer mechanical degradation via two types C – C bond cleavage: (i) homogeneous and (ii) heterogeneous. Generation process of polymer mechano-anions is combined with the polymer triboelectricity phenomenon caused by mechanical scission of the polymer main chain on a friction surface.

It is generally known that the friction between dielectrics results in the buildup of electric charge. The energy of electrons emitted from polymers amounts to scores of keV, with the emission itself being a lengthy, slowly decaying process; such electrons are called mechano-electrons. Most recent paper [67] reports (see also references in [67]) the mechano-emission arises as a result of the ionization of surface traps at the expense of the energy which is released in the annihilation of the defects which are formed during cleavage; the slow electrons are accelerated in the field of negatively charged segment of the freshly cleaved surface. Slow electrons appear upon the ionization of surface traps. The energy of the electrons was evaluated from the deviation of the electron beam in a magnetic field and by measuring their passage through obstacles.

At this point the following question is asked. Is here the only mechanism involved that electron transfers from the mechano-anion (R$^{\ominus}$) produced via heterogeneous bond cleavage as suggested by Sakaguchi et al. [68]? The electron transfer from R$^{\ominus}$ was documented by the following reaction in dark with tetracyanoethylene (TCNE) electron scavenger as follows:

$$R^{-} + \text{TCNE (mixing in the dark)} \longrightarrow R^{\bullet} + \text{TCNE}^{-\bullet} \quad (7)$$

Figure 18 summarizes all the process steps, where •R1 and •R2 relate to free radicals of homogeneous C–C bond cleavage. R^- and R^+ relate to anion and positive ion species of the heterogeneous polymer bond scission. Radical R• concerns electron transfer from the negative ion R^- (anion) to TCNE either during mixing or photo irradiation.

$$R1 - R2 \rightarrow {}^{\bullet}R1 + {}^{\bullet}R2 \quad \text{Homogeneous}$$
$$R - R \rightarrow R^- + R^+ \quad \text{Heterogeneous}$$

$$R^- + TCNE \xrightarrow{\text{mixing in the dark}} R^{\bullet} + TCNE^{\bullet -}$$

$$R^- + TCNE \xrightarrow{\text{photoirr}} R^{\bullet} + TCNE^{\bullet -}$$

Figure 18. Illustration and suggested evidence for heterogeneous C–C bond cleavage based on work [68].

The electron transfer reaction in the dark at 77K is promoted by physical mixing of fractured sample in the vibration glass ball mill and, after milling; the fractured sample was dropped into the ESR sample tube under vacuum in the dark at 77K and photo irradiated using an IR lamp with a glass filter corresponding to visible light [68]. Thus, everything is clear for the irradiation effect due to electron detachment. On the other hand, for reaction **(7)** here an alternative mechanism is proposed, as detailed in Fig. 19.

Polymer + TCNE ***(mixing in the dark)*** **→**

R^+ ***(positively charged site)*** + $R^{\bullet}$ ***(free radical)*** + **e** ***(emited electron)***

$$TCNE + e \rightarrow \underline{TCNE}^{\bullet -}$$

and thus, according to the NIRAM approach,

the electron attachment to

TCNE

replaces the heterogeneous C — C bond cleavage

$$R - R \rightarrow R^- + R^+$$

Figure 19. Application of the NIRAM approach to account for the mechano-anion-radical generation

Actually, generation of mechano-anion-radicals relates strictly to triboelectricity and is reviewed in reference [68]. It cites a wide range of papers which aimed at finding evidence for both heterogeneous C-C bond scission and polymer triboelectricity. The review strongly suggests production, eg. mechano-anions induced by mechanical fracture of PCV. EPR studies provide evidence TCNE radical-anion generation. However, the formulation of a satisfactory theory to account for the triboelectricity of polymers has yet to be established. The same is due to clear evidence for heterogeneous C – C bond scission. An extension of the NIRAM approach to better understand the mechano-anion-radicals allows considering tribochemistry as a branch of mechanochemistry.

Triboelectricity or contact electrification of materials is a very complex phenomenon. According to [69] the first studies on contact electrification were carried out over 2500 years ago, when experiments showed that rubbing amber and wool caused the two materials to become oppositely charged. Our scientific understanding of contact electrification has not progressed too much and, it is still not known what species is being transferred between the wool and amber to generate the charge, and how rubbing influences the process [70]. That review paper concludes with a discussion that virtually all questions involving electrostatics are in fact open ones and the size of existing particles is of special importance. It was assumed and partly evidenced that small particles charge negatively and the larger particles should charge positively. Fig. 20 illustrates the effect of contact geometry for contact charging of bulk-scale surfaces of identical insulator materials [70] and, some works referred in it.

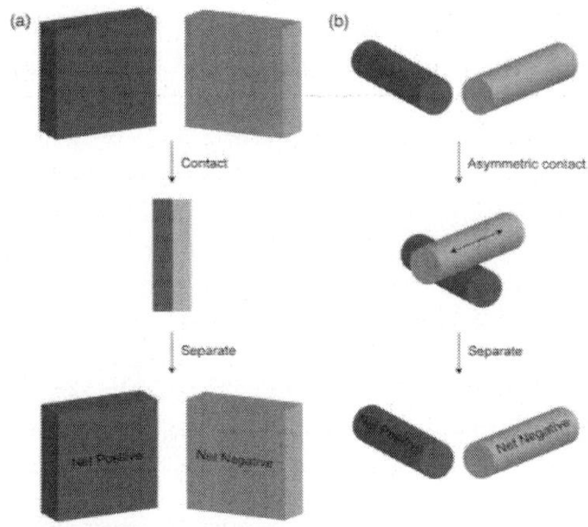

Figure 20. Contact between two material surfaces in a symmetric fashion resulting in a net positive charge on one surface and a net negative in an apparently random direction (*a*); contact between two materials' surfaces in an asymmetric fashion usually results in a net negative charge on the surface with the smaller contacting area and net positive charge on the surface with the larger contacting area (*b*). Figure taken from Reference [70].

6. Present knowledge on mechanochemistry and tribochemistry

6.1. Brief introduction

Particularly important research and practical application of mechanochemistry is well reflected in INCOME (International Conference on Mechanochemistry) series of special meetings initiated in 1993 by International Mechanochemistry Association (IMA). IMA is the associate member of International Union of Pure and Applied Chemists (IUPAC). These conferences regularly serve as a common platform to bring together all stakeholders from academia, research and development organisations, along with industry to foster the growth of the discipline [71]. The first INCOME meeting was held in Slovakia (1993). Participants from 25 countries of 4 continents took part at the meeting. This international forum was preceded by eleven All-Union Symposia on Mechanochemistry and Mechanoemission.

INCOME 2011 focused on mechanochemistry and mechanical alloying, was held in Herceg Novi, Montenegro. The conference aimed at providing a forum for presentation of new results, identification of current research and development trends along with future directions and, promoted interactions dealing with various aspects of the discipline. Presented research papers concerned both mechanochemistry and tribochemistry. By and large, they demonstrated the progress of studies on the chemical and physicochemical processes proceeding in solids under mechanical action.

6.2. Major recent review papers

Some review papers have already been described in previous sections. They specifically relate to mechanochemistry [46,66] and/or tribochemistry [6,11,29]. The most recent extensive critical review paper [72] aims at providing a broad but digestible overview of mechanochemical synthesis, that is reactions conducted by grinding solid reactants together with no or minimal solvent. This critical review includes over 300 references focused on the historical development, mechanistic aspects, limitations and opportunities of mechanochemical synthesis. It emphasizes that although mechanochemistry has historically been a sideline approach to synthesis, presently it may move into the mainstream because it is increasingly apparent that it can be practical, and even advantageous [72]. This is because it provides the opportunities for developing more sustainable technologies. Additionally, synthesis of metal organic frameworks (MOFs) has to be mentioned here, as it have become presently one of the most intensely researched areas of materials chemistry and, it might relate to both mechanochemistry and tribochemistry.

An earlier detailed review paper generally aimed at mechanical activating of covalent bonds [73], also considers mechanochemistry of crystals, metals, alloys, and polymers. The latter is related to already discussed in detail heterogeneous C – C bonding splitting in polymers [68] (see Section 5.3). By the employing of mechanochemical solid-state reaction, Fujiwara and Komatsu synthesized a novel C_{60} dimer linked by a silicon bridge and a single bond [74]. The novel C_{60} dimer was synthesized using a high-speed vibration milling (HSVM) technique and, the product obtained was fully characterized by a wide variety of sophisticated analytical techniques. The electronic interaction between the two C_{60} cages was evidenced by the electrochemical method. For instance, in these dimers, the two fullerene cages are connected by sharing a cyclobutane ring. the mechanochemical solid-state reaction of C_{60} with alkyl or aryl halide in the presence of alkali metals was found to cause alkylation or arylation of C_{60} possibly by the intermediacy of the C_{60} radical anion [74]. The mechanochemical reaction of the fullerene with dichlorodiphenylsilane (Ph2SiCl2) and lithium (Li) metal, under the solvent free conditions, allowed synthesizing the C_{60} dimer fused with a silacyclopentane ring, as shown below [74].

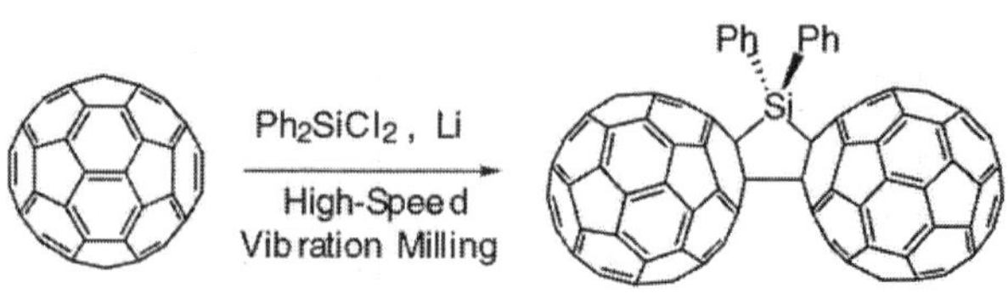

6.3. Practical aspects of mechanochemistry

Very detailed review publications [72-73] include a wide range of industrial applications of mechanochemistry and underline the need for sustainability brought about by the Kyoto Treaty and the increasing global demand for products leading to an increase in sustainable manufacturing processes. Such processes and/or new technologies can lower environmental demands. Improved sustainability can be realized in the form of reduced energy use, less organic solvents, better selectivity and reduced waste. Recent well specified study [75] presents several advantages of mechanochemical technology, such as simple process, ecological safety and the possibility of obtaining a product in the metastable state. Thus, it gives an overall review of the mechanochemistry applications in waste management. Interestingly to note that based on that study, the modification of fly ash and asbestos containing wastes (ACWs) can be achieved by mechanochemical technology (MCT). Finally, MCT provides with a prospective application in pollution remediation and also waste management. Hazardous metal oxides can be transformed into easily recyclable sulfide by mechanochemical sulfidization; the waste plastics and rubbers, which are usually very difficult to be recycled, can also be recycled by mechanochemical technology [75]. At this point the importance of waste-free mechanochemical synthesis and production should also be noted [5].

Another study [76] demonstrates that the mechanochemical treatment of an expired drug as ibufren, may be very useful and non-polluting way to change the bioactive molecule of a pharmaceutical formulation into non-toxic product. In the degradation mechanism two major steps are considered. The first one in the degradation process that leads to the detoxification of the drug, is induced by the presence of a co-reagent aluminum hydroxide. The second step relates to oxidative decarboxilation to an intermediate stable benzyl radical which, reacting in the presence of oxygen during milling, produces several products. The number of the generated products is controlled by the milling time. Thereafter, mechanochemical treatments are environmentally sound for general approach to detoxify chemical-pharmaceutical waste. Thus, in this case a type of molecular mechanolysis without solvent or usual thermal contribution can be considered [76].

Extension of mechanochemistry practical applications is broadly described by Todres in the book on both mechanochemistry and tribochemistry [77]. The specificity of the Todres' book is that it discusses chemical reactivity of organic molecules mechanically treated separately or together and, describes mechanochemically initiated polymerization, depolymerization and mechanolysis. It also considers lubricant design and lubricity process. Organic solids/materials are very liable to attrition representing wear caused by friction (tribos) and therefore relating to tribochemistry. This is evidenced by considering detailed organic reactions with lubricating layers.

6.4. How mechanochemistry is entangled with tribochemistry

6.4.1. Selected books and review publications

Mechanochemistry comprises a wide branch of the reactivity research of solids activated by mechanical action. It also deals with mechanisms of solid-phase reactions, aiming at making new materials by non-traditional methods. Nowadays, it is an established field of materials science and solid-state chemistry [58]. The importance of mechanochemistry is evidenced by several books clearly presenting its scientific and practical achievements [46,78-80].

The latter book [80] is on powder technology and deals with a variety of particles, from submicrometer to large grains and gravels and from liquid mist or droplets to bubbles, as well as solid particles and aggregates. As it considers with two-phase and three-phase mixtures, it relates to work in different science and engineering fields and, thereby demonstrate a bridge between pure mechanochemistry, inorganic chemistry and/or technology, organic chemistry and/or technology and, tribochemistry in its broad meaning. Thereby, it evidences simultaneously how mechanochemistry is complex and entangled with tribochemistry. On the other hand, a lot of natural and artificial phenomena encountered in our daily lives may be accounted for by knowledge of powder technology.

Some books are related to tribology [15,28] or tribochemistry [7-8], however they also consider mechanochemistry. Another one tends to tribochemistry via organic mechanochemistry [76]. The same is due to review papers [13, 16-17] and book chapters [48-50].

In summary, presently both mechanochemistry and tribochemistry appear to be a science with a sound theoretical foundation. Their major benefits include lower reaction temperatures and increased reaction rate. As a consequence, processing of materials can be performed in simpler and less expensive reactors during shorter reaction times.

6.4.2. Mechanochemistry vs. tribochemistry

It is known that some chemical substances react differently when exposed to mechanical and thermal energy. The term mechanochemistry was coined by Oswald for the corresponding branch of chemical physics. In this sense, mechanochemistry should be considered along with thermochemistry, electrochemistry, photochemistry, sonochemistry, chemistry of high pressures, shock waves, or microwave effects [76]. In fact, organic solids are very liable to attrition; inorganic materials appear to be more resistant [81] and the attrition represents wear caused by rubbing or friction which already relates to tribochemistry.

At this point we need to come back to the Todres book [77], which outlines the main regularities governing transformations of mechanically activated organic compounds and discusses physical processes that affect these transformations. Some chapters concern more specifically the mechanically induced reactions of organic synthesis and the chemical transformations of organic participants of boundary lubrication. As already indicated, the mechanically induced synthesis of the desired organic compounds is advantageous in the sense of rates of formation.

Summarily, this is preferred over expenses required for mechanical activation (say, for the electricity spent to rotate a mill); chemical forces arise from summation of excess enthalpy of individual participants and that of chemical interaction. Thus, such chemical processes should be described in terms of the tribochemistry action mechanisms.

6.4.3. Similarities between mechanochemistry and tribochemistry

Definitions of mechanochemistry and tribochemistry, formulated 50 years ago [7,8], are generally accepted and, the following definitions have been selected for this Chapter. *Mechanochemistry is a branch of chemistry which is concerned with chemical and physicochemical transformations of substances in all states of aggregation produced by the effect of mechanical energy. Tribochemistry is a branch of chemistry dealing with the chemical and physico-chemical changes of solids due to the influence of mechanical energy.*

Mechanochemistry is the science field that deals with athermal or ultra-fast chemical reactions between solids or between solids and surrounding gaseous or liquid molecules under mechanical forces. Another approach defines mechanochemistry as the branch of solid state chemistry where intramolecular bonds are mechanically broken [5]. Actually, the same is due to tribochemistry. Both mechanochemical and tribochemical reactions are also distinct from those of themochemical reactions. The same is due to heterogeneous catalysis (HetCat) and tribocatalysis [11]. Triboemission, triboplasma, and NIRAM-HSAB approach consider electronic effect of on mechanochemistry and tribochemistry. To initiate thermochemical reactions heat should be supplied. Thus, it seems convenient to consider heat evolution also in electronic terms. Additionally, it is suggested to consider tribochemistry as a subset of mechanochemistry.

There is an overlap of disciplines based on chemical reactions initiated by the mechanical action. Not considering any heat loss, the difference is controlled by *energy stored* in the system. Looking at the mechanical work plane proposed in [15], various portions of the work (power) include: input power, use-output power, loss-output energy rate, and a stored energy (thermal energy transformed from mechanical work). The energy stored (excess energy) serves as the origin of enhanced reactivity of solids.

6.4.4. Differences between mechanochemistry and tribochemistry

Taking into account that tribochemistry is the subset of mechanochemistry, major differences between mechanochemistry and tribochemistry is formal, as compared with chemical physics and physical chemistry. There are many physical processes related to the wearing and mechanisms. These mechanisms are often connected with tribochemical reactions. The nature of the tribochemical film is the key to understanding the mechano-chemical processes that give rise to chemical films separating friction solid elements.

Significant evidence for the above suggestion comes from book [77]. That book aimed at correlating mechanical actions on organic substances with the molecular events triggert by these actions. The term *organic mechanochemistry,* defined as *convertion of mechanical energy into the driving force for molecular oe structural phase trtansitions,* was introduced. Importantly, Todres in his book [77] emphasizes that chemical engineering needed *inorganic mechanochemistry* be addressed first. *Organic mechanochemistry* has been in its infancy for a long time. The specificity of the Todres' book is that it discusses chemical reactivity of organic molecules mechanically treated separately or together and, describes mechanochemically initiated polymerization, depolymerization and mechanolysis. It also considers lubricant design and lubricity process and, differentiates mechanochemical publications from tribochemical ones.

At this point, it is proposed to assign *inorganic mechanochemistry* to *mechanochemistry* and, *organic mechanochemistry* to *tribochemistry.* Thus, major differences between mechanochemistry and tribochemistry have a formal character.

6.4.5. Common denominators of mechanochemistry and tribochemistry

Considering that the excess energy (energy stored in a sytem) serves as the origin of enhanced reactivity of materials under mechanical treating, their common denominators broadly encompass triboemission and triboplasma processes.

In the present Chapter understanding, there is only a formal overlap of these disciplines in terms of inorganic and organic chemistry. Accordingly, *mechanochemistry* is the *inorganic mechanochemistry* and, *tribochemistry* is the *organic mechanochemistry*. Tribochemistry is considered as a subset of mechanochemistry.

The major denominatot for inorganic mechanochemistry, organic mechanochemistry, catalysis and tribocatalysis is the NIRAM-HSAB approach, because it allows accounting for most of very specific tribological findings. It also demonstrates how organic chemistry is changing via physical chemistry, chemical physics and nowadays by contributing to tribochemistry. Book chapter [82] details physical and chemical phenomena concerning the tribochemistry discipline.

7. Conclusions

Mechanically initiated chemical reactions in solids are not new and presently there are many practical and theoretical achievements. Since mechanically initiated chemical reactions in solids have not received adequate attention yet, first of all a better understanding between researchers dealing with mechanochemistry and tribochemistry is needed. It was a major goal of this Chapter.

The new approach to mechanically initiated chemical reactions shows that the NIRAM-HSAB theory is important in accounting for most of very specific tribological findings and, also allows to account for the first mechanochemical reactions, for example

$$\mathbf{HgS + Cu \longrightarrow Hg + CuS} \tag{2}$$

or modify the mechanism of heterogeneous C - C bond splitting in polymers.

Author details

Czesław Kajdas
Automotive Industry Institute PIMOT, Warsaw
Warsaw University of Technology, Institute of Chemistry, Płock, Poland

Acknowledgement

The author wishes to acknowledge PIMOT for this project financial support and, Krzysztof Kowalczyk for his assistance in preparing figures.

8. References

[1] Carey-Lea M. On Endothermic Reactions Effected by Mechanical Force. Philosophical Magazine 1893, 36, 350-351.

[2] Takacs L. M. Carey Lea. The Father of Mechanochemistry. Bulletin for the History of Chemistry 2003, 28(1), 26-34.

[3] Takacs L. M. Carey Lea, the First Mechanochemists. Journal of Materials Science 2003, 39(16-17), 4987-4993. See also: Takacs L. The First Documented Mechanochemical Reaction? Journal of Metals 2000; (Jan issue) 12-13.

[4] Vick B., Furey M. J., Kajdas C. An Examination of Thermionic Emission Due to Frictionally Generated Temperatures, Tribology Letters 2002, 13(2), 147-153.

[5] Kaupp G, Waste-Free Synthesis and Production all Across Chemistry with the Benefit of Self- Assembled Crystal Packings. Journal of Physical Organic Chemistry 2008, 21(7-8), 630-643. ISSN: 08943230; DOI: 10.1002/poc.1340.

[6] Kaupp G. Mechanochemistry: the Varied Applications of Mechanical Bond-Breaking, The Royal Society of Chemistry, CrystEngComm 2009, 11, 388–403. DOI: 10.1039/b810822f ...

[7] Thiessen P.D., Meyer K., Heinicke G. Grundlagen der Tribochemie. Berlin: Akademie-Verlag, 1966.

[8] Heinicke G. Tribochemistry. Berlin: Academy-Verlag, 1984.

[9] Hardy W. Collected Works. Cambridge: University Press, 1936.

[10] Campbell, W.F. Boundary Lubrication. In: Ling F.F., Klaus E.E., Fein R.S. (eds) Boundary Lubrication. An Appraisal of World Literature. New York: ASME; 1969.pp87-117.

[11] Kajdas C., Hiratsuka K. Tribochemistry, Tribocatalysis, and the Negative-Ion-Radical Action Mechanism. Proceedings of the Institution of Mechanical Engineers Part J: Journal of Engineering Tribology 2009; 223 (6) 827-848. DOI: 10.1243/13506501JET514

[12] Bond G.C. Heterogeneous Catalysis. Principles and Applications. Oxford: Clarendon Press, 1987.

[13] Hsu S.M., Zhang J., Yin Z. The Nature and Origin of Tribochemistry. Tribology Letters 2002, 13(2) 131-139.

[14]http://books.google.pl/books?id=8lsEsGe18t8C&pg=PA587&lpg=PA587&dq=definition+of+tribochemistry&source=bl&ots=UFdquGVHoK&sig=HQoet9rQjpUgUb7c0b97TzXHIx8&hl=pl&sa=X&ei=WLyXT9HJEsrf4QTDyqDFBg&sqi=2&ved=0CDAQ6AEwAQ#v=onepage&q=definition%20of%20tribochemistry&f=false

[15] Czichos H. Tribology a Systems Approach to the Science and Technology of Friction, Lubrication and Wear. Amsterdam: Elsevier; 1978.

[16] Butyagin P.Yu. Kinetics and Nature of Mechanochemical Reactions. Russian Chemical Reviews 1971, 140 (11) 901-915.

[17] Guha D, Roy Choudhuri S.K. The effect of surface roughness on the temperature at the contact between sliding bodies. Wear 1996, 197 (1-2) 63–73. DOI: 10.1016/j.bbr.2011.03.031

[18] Kajdas C., Harvey S.S.K., Wilusz E. Encyclopedia of Tribology. Amsterdam: Elsevier; 1990.

[19] Kalin M. Influence of Flash Temperatures on the Tribological Behavior in Low-Speed Sliding: a Review. Materials Science and Engineering A 374 (2004) 390–397. DOI:10.1016/j.msea.2004.03.031

[20] Weick BL, Furey MJ, Vick B. Surface Temperatures Generated with Ceramic Materials in Oscillating/Fretting Contact. Transactions of the ASME, Journal of Tribology 1994, 116 (4) 260-267.

[21] Archard JR. The Temperature of Rubbing Surfaces. Wear 1959, 2(6) 438-455.

[22] Archard JR., Rowntree RA. The Temperature of Rubbing bodies.Part 2. The Distributions of Temperatures. Wear 1988, 128(1) 1-17.

[23] Marscher WD. A Critical Evaluation of the Flash Temperature Concept. ASLE Transactions 1982, 25(2) 157-174. DOI: 10.1080/05698198208983077

[24] Blok H. Theoretical Study of Temperature Rises of Actual Contact under Oiliness Lubricating Conditions. Proceedings of General Discussion on Lubrication. London: Instn Mech Engr; 1937 vol. 2. p. 222–35.

[25] Blok H. The flash temperature concept. Wear 1963, 6(6) 483-494. DOI.org/10.1016/0043-1648(63)90283-7

[26] Archard JF. Contact and Rubbing of Flat Surfaces. Journal of Applied Physics 1953; 24:981–988.

[27] Carslaw HS, Jaeger JC. Conduction of heat in solids. Oxford: Oxford University Press; 1959.

[28] Soom A, Serpe CI, Dargush GF. Thermomechanics of Sliding Contact. Fundamentals of Tribology and Bridging the Gap between the Macro and Micro/Nanoscales. NATO Science Series. II. Mathematics, Physics and Chemistry, vol. 10. Dordrecht: Kluver; 2001 p. 467–85.

[29] Kajdas CK. Importance of the Triboemission Process for Tribochemical Reaction. Tribology International 2005, 38 (3) 337–353. DOI:10.1016/j.triboint.2004.08.017

[30] Vick B, Furey MJ. A Basic Theoretical Study of the Temperature Rise in Sliding Contact with Multiple Contacts. Nordtrib' proceedings, vol. 2. Porvoo, Finland: VTT; 2000 p. 389–98.

[31] Vick B., Furey MJ, Foo SJ.Boundary Element Thermal Analysis of Sliding Contact. Numerical Heat Transfer, Part A: Applications: An International Journal of Computation and Methodology 1991, 20(1) 19-40. DOI:10.1080/10407789108944807

[32] Tagawa M., Mori M., Ohmae N., Umeno M., Takenobu S. Tribo- and Photo-Stimulated Exoelectron Emission from Graphite. Tribology International 1993, 26(1) 57-60. ISSN: 0301679X

[33] Momose Y, Iwashita M. Surface Analysis of Metals using Tribostimulated Electron Emission. Surface and Interface Analysis 2004, 36(8) 1241–1245. DOI: 10.1002/sia.1885

[34] Nevschupa RA. Triboemission: An Attempt of Developing a Generalized Classification. In: Tribology Science and Application, Herman MA (Ed.), CUN PAN, Warsaw; 2004 p. 11–25

[35] Nakayama K, Suzuki N, Hashimoto H. Triboemission of Charged Particles and Photons from Solid Surfaces during Frictional Damage. Journal of Physics D: Applied Physics 1992; 25(2) 303–308. DOI:10.1088/0022-3727/25/2/027

[36] Kim MW, Langford SC, Dickinson JT. Electron and Photon Emission Accompanying the Abrasion of MgO with Diamond. Tribology Letters 1 1995, 1(2-3) 147-157. DOI: 10.1007/BF00209770

[37] Molina GJ, Furey MJ, Ritter AL, Kajdas C. Triboemission from Alumina, Single Crystal Sapphire, and Aluminum. Wear 2001; 249(3-4) 214–219. PII: S0043-1648(01)00568-3

[38] Molina GJ, Furey MJ, Vick B, Ritter AL, Kajdas C. Triboemission from the Sliding Contact
of Alumina Systems. Proceedings of the Second World Tribology Congress, Vienna, Austria 2001.

[39] Molina GJ, Triboemission from Ceramics: Charge Intensity and Energy Distribution Characterizations. PhD Dissertation, Deptartment of Mechanical Engineering, Virginia Polytechnic Institute and State University, Blacksburg; 2000.

[40] Kajdas C. On a Negative-Ion Concept of EP Action of Organo-Sulfur Compounds. ASLE Transaction 1983; 28(1) 21–30.

[41] Sato N, Seo M. Chemically Stimulated Exo-Emission from a Silver Catalyst. Nature 1967; 216(Oct) 361-362. DOI:10.1038/216361a0

[42] Ramsey JA. The Emission of Electrons from Aluminum Abraded in Atmospheres of Air, Oxygen, Nitrogen and Water Vapor. Surface Science 1967, 8(3) 313–322. DOI.org/10.1016/0039-6028(67)90114-8

[43] Varentsov EA, Khrustalev YA. Mechanoemission and Mechanochemistry of Molecular Organic Crystals. Russian Chemical Reviews 1995; 64(8) 783-797. IOP.org/0036-021X/64/8/R06

[44] Khrustalev YA. Electric Phenomena on the Rupturing of Adhesive Contact and Failure of Solids: Development Stages from Gas Discharge to Cold Nuclear Fusion. Colloids and Surfaces A: PhysicoChemical and Engineering Aspects 1993, 79(1) 51–63. DOI.org/10.1016/0927-7757(93)80159-C

[45]. Thiessen KP, Sieber K. Energetische Randbedingungen tribochemischer Prozesse. Teil 3. Chemie Leipzig 1979, 260(3) 410-416.

[46] Baláž P. Mechanochemistry in Nanoscience and Minerals Engineering. Berlin: Springer-Verlag,
2008.http://books.google.pl/books?id=FldqbSffUMgC&pg=PA1&lpg=PA1&dq=Bal%C3%A1%C5%BE+P.+Mechanochemistry+in+Nanoscience+and+Minerals+Engineering.+Berlin:+Springer-Verlag,+2008

[47] Pawlak Z. Tribochemistry of Lubricating Oils. Amsterdam: Elsevier; 2003.

[48] Tysoe TT, Kotvis PT. Surface Chemistry of Extreme Pressure Lubricant Addtitives. In Surface modification and mechanisms (Eds G. E. Totten and H. Liang), 2004, ch. 10, pp. 299–351 (Marcel Dekker, Inc., New York, Basel).

[49] Vižintin J. Additive Reaction Mechanisms. In Surface modification and mechanisms (Eds G. E. Totten and H. Liang), 2004, ch. 9, pp. 243–298. (Marcel Dekker, Inc., New York, Basel).

[50] Buyanovsky IA, Zimaida VI, Zaslavsky RN. Tribochemistry of Boundary Lubrication Processes. In Surface modification and mechanisms (Eds G. E. Totten and H. Liang), 2004, ch. 11, pp. 353-404. (Marcel Dekker, Inc., New York, Basel).

[51] Kajdas C. Tribochemistry. In Surface modification and mechanisms (Eds G. E. Totten and H. Liang), 2004, ch. 6, pp. 99-164. (Marcel Dekker, Inc., New York, Basel).

[52] M. Majzner, Kajdas C. Reactions of Carboxylic Acids under Boundary Friction Conditions. Tribologia 2003; 34(1) 63–80 (in Polish).

[53] Fischer DA, Hu ZS, Hsu SM. Tribochemical and Thermochemical Reactions of Stearic Acid on Copper Surfaces in Air as Measured by Ultrasoft X-Ray Absorption Spectroscopy.Tribology X-Ray Absorption Spectroscopy Tribology Letters 1997; 3(10) 35–40. DOI: 10.1023/A:1019109407863

[54] Makowska M, Kajdas C, Gradkowski M. Interactions of Hexadecane with 52100 Steel Surface under Friction Conditions. Tribology Letters 2002;13(2) 65–70. http://www.ingentaconnect.com/content/klu/tril/2002/00000013/00000002/00377942

[55] Kajdas C, Shuga'a AK. Investigation of AW Properties and Tribochemical Reactions of Esters of Palmitic Acid and Aliphatic Alcohols in the Steel-on-Steel System. Tribologia 1998; 29(xx) 389–402 (in Polish).

[56] Kajdas C. Hydrolysis. In Surface modification and mechanisms (Eds G. E. Totten and H. Liang), 2004, ch. 8, pp. 203-241. (Marcel Dekker, Inc., New York, Basel).

[57] von Ardenne M, Steinfelder K, Tuemmler R. Elektronenanlagerungs-Massenspektrographie organischer Substanzen. Berlin: Springer; 1971.

[58] Gilman JJ. Mechanochemistry. Science 1996; 274 (5284) 65. DOI: 10.1126/science.274.5284.65

[59] Burdett JK. Chemical Bonding in Solids. New York: Oxford University Press; 1995. http://books.google.pl/books/about/Chemical_Bonding_in_Solids.html?id=Z-C0QENZVTIC&redir_esc=y

[60] Dante RC, Kajdas CK. A Review and a Fundamental Theory of Silicon Nitride Tribochemistry. Wear 2012; 288(5) 27–38. http://dx.doi.org/10.1016/j.wear.2012.03.001

[61] Gates RS, Hsu SM. Boundary Lubrication of Silicon Nitride, National Institute of Standards and Technology (NIST) Special Publication 876. NIST, Gaithersburg, MD 20899-0001, USA, February 1995, pp. (i)–(xv) and 1–379.

[62] Gates RS, Hsu SM. Silicon Nitride Boundary Lubrication: Lubrication mechanism of Alcohols, Tribology Transactions 1995; 38(3) 645–653. ISSN 0569-8197

[63] Tsutsumi T, Hiratsuka K, Ohta K, Kajdas C. Activation Energy of Tribocatalysic Oxidation of Ethylene. P17 at the International Tribochemistry Conference Hagi, Japan 2012.

[64] Molchanov VV, Buyanov RA. Mechanochemistry of Catalysts. Russian Chemical Reviews 2000; 69 (5) 435- 450. DOI 10.1070/RC2000v069n05ABEH000555

[65] Hill J. Theophrastus' History of Stones. London 1774; p. 235.

[66] Boldyrev VV, Tkacova K. Mechanochemistry of Solids: Past, Present, and Prospects. Journal of Materlials. Synthesis and Processing 2000; 8(3-4) 121-132.

[67] Chandra BP, Patel NL, Rahangdale SS, Patel RP, Patle VK. Characteristics of the Fast Electron Emission Produced During the Cleavage of Crystals. PRAMANA Journal of Physics. Indian Academy of Sciences 2003; 60(1) 109–122.

[68] Sakaguchi M, Miwa Y, Hara S, Sugino Y, Yamamoto K, Shimada S. Triboelectricity in Polymers: Effects of the Ionic Nature of Carbon–Carbon Bonds in the Polymer Main Chain on Charge due to Yield of Mechano-Anions Produced by Heterogeneous Scission of the Carbon–Carbon Bond by Mechanical Fracture. Journal of Electrostatics 2004; 62(1) 35-50. http://dx.doi.org/10.1016/j.elstat.2004.04.003

[69] O'Grade PF. Thales of Miletus: The Beginnings of Western Science and Philosophy. Aldershot, UK: Ashgate 2002.

[70] Lacks DJ, Sankaran RM. Contact Eectrification of Isulating Materials. Journal of Physics D: Applied Physics 2011; 44, 453001 (15pp), DOI:10.1088/0022-3727/44/45/453001

[71] INCOME Conference 2011: http://www.mrs-serbia.org.rs/income2011/income2011.html

[72] James SL et al. Mechanochemistry: Opportunities for New and Cleaner Synthesis. Chemical Society Reviews of the Royal Society of Chemistry 2012; 41, 413–447. DOI: 10.1039/c1cs15171a

[73] Beyer KB, Clausen-Schaumann H. Mechanochemistry: the Mechanical Activation of Covalent Bonds. Chemical Reviews 2005; 105(8) 2921-2948. DOI:10.1021/cr030697h

[74] Fujiwara K, Komatsu K. Mechanochemical Synthesis of a Novel C60 Dimer Connected by a Silicon Bridge and a Single Bond. Organic Letters 2002; 4(6) 1039-1041. DOI:10.1021/ol025630f.

[75] Guo X, Xiang D, Duan G, Mou P. A Review of Mechanochemistry Applications in Waste Management. Waste Management 2010; 30, 4–10.

[76] Andini S, Bolognese A, Formisano D, Manfra M, Montagnardo F, Santoro. L. Mechanochemistry of ibuprofen pharmaceutical. Chemosphere 2012: article in press; Available online 1 April 2012: http://dx.doi.org/10.1016/j.chemosphere.2012.03.025

[77] Todres ZV. Organic Mechanochemistry and its Practical Applications. Boca Raton: CRS Taylor & Francis Group, LLC; 2006.

[78] Gutman EM. Mechanochemistry of Materials. Cambridge: Cambridge International Science Publishing; 1998.

[79] Baláž P. Extractive Metallurgy of Activated Minerals (Process Metallurgy). Amsterdam: Elsevier Science; 2000.

[80] Powder Technology Handbook. Masuda H, Higashitani K, Yoshida H, (eds). Boca Raton: CRS Taylor & Francis Group, LLC; 2006.

[81] Bravi, M., Di Cave, S., Mazzarotta, B., Verdone, N. Relating the Attrition Behaviour of Crystals in a Stirred Vessel to their Mechanical Properties. Chemical Engineering Journal 2003; 94(3) 223-229. DOI:10.1016/S1385-8947(03)00053-67-353.

[82] Kajdas C. Physical and chemical phenomena related to tribochemistry. In Advances in contact mechanics: Implications for materials science, engineering and biology. (Eds R. Buzio and U.Valbusa), 2006, ch. 12, pp. 383-412. (Transworld Research Network, Kerala, India).